Carolina Zeferino

Prospects for Family Farming

Carolina Zeferino

Prospects for Family Farming

A study of milk production in 2017

ScienciaScripts

Imprint
Any brand names and product names mentioned in this book are subject to trademark, brand or patent protection and are trademarks or registered trademarks of their respective holders. The use of brand names, product names, common names, trade names, product descriptions etc. even without a particular marking in this work is in no way to be construed to mean that such names may be regarded as unrestricted in respect of trademark and brand protection legislation and could thus be used by anyone.

Cover image: www.ingimage.com

This book is a translation from the original published under ISBN 978-613-9-65578-6.

Publisher:
Sciencia Scripts
is a trademark of
Dodo Books Indian Ocean Ltd. and OmniScriptum S.R.L publishing group

120 High Road, East Finchley, London, N2 9ED, United Kingdom
Str. Armeneasca 28/1, office 1, Chisinau MD-2012, Republic of Moldova, Europe
Printed at: see last page
ISBN: 978-620-7-78430-1

SUMMARY

ACKNOWLEDGMENTS 3
SUMMARY 4
1. INTRODUCTION 5
2. LITERATURE review 7
3. METHODOLOGICAL PROCEDURES 23
4. CONTEXTUALIZATION OF THE MUNICIPALITIES 25
5. CONCLUSION 60
REFERENCES 62
APPENDIX 1: INTERVIEW WITH DAIRY FARMERS 67

I would like to dedicate this work to my family, especially my parents, who have worked so hard to achieve this feat. To all my close friends and family for their support and contribution to this research.

ACKNOWLEDGMENTS

The realization of this work, in addition to all the commitment and dedication on my part, counted on the participation of special people who were part of this trajectory of studies, who made it possible to make this dream come true.

First of all, I thank God for the grace to live, and even in the face of certain difficulties and setbacks, he has given me the health, calm and perseverance not to lose heart.

To all my family, especially my parents Nilza and Vanderlei Zeferino, who were my main foundations, supporting me through a difficult stage, sparing no effort to support me, both with welcoming words and financially to reach this moment.

To Douglas Kauê Corrêa, for his companionship, attention, help and support, which were extremely important throughout this research.

To UNIDAVI for the space made available and its structure, together with the professors who taught my course, always fulfilling their role as educators. In particular, my advisor Anielle Gonçalves de Oliveira for her willingness to provide corrections and support, as well as Professor Jeancalo Visentainer, who was always willing to answer my questions about the methodological requirements, welcoming me with attention, both of which together were essential to my academic development for this monograph.

To Epagri, especially to the agronomist Leliani Valéria de Souza, who welcomed me with attention, clearing up my doubts about the choice of municipalities, helping whenever necessary to carry out this work.

The Marcelo, Will, Grahl, Vanderlinde and Norenberg families, who were willing to take part in the interviews and were welcomed into their homes, were available until the end of the work to answer any questions.

And everyone who contributed with their actions, prayers and positive thoughts to the realization of this research.

SUMMARY

Farming is the main activity for human beings, both for subsistence and to develop a given region through a productive unit. As it is a demanding and decisive job, it has gone through a process of giving up, initiating a predominant rural exodus among small producer families, in the face of poor working conditions, as well as a lack of incentives. Over the years, innovation and technology have reached the agricultural sector, resulting in less manual and costly work in the production process. This scenario has made it easier to handle, but it has also created some discomfort for farmers who used to produce in a cultural way, giving the space to their children. The aim of this study is to analyze young people's prospects of succession in dairy farming, as well as the influence of innovation and technology, by interviewing families. Five properties were visited, presenting the prospects for family succession, together with the benefits that technology and innovation have made available to them. The methodology used was firstly bibliographical research to develop the theoretical framework. This was followed by a field study of the owners interviewed. As with other activities, dairy farming faces a number of challenges, but the families have found a better quality of life and working conditions. Through conversation, it was possible to see how much the knowledge acquired through courses offered by rural programs, technical courses and even undergraduate degrees, contributes to improving agricultural activity, helping to adapt technology to their environment. It can be seen that the technological methods applied have considerably increased the permanence of young people who used to work outside the farm, helping them to stay in milk production and the rural environment, and in turn reducing the concentration of the rural exodus.

Keywords: Family farming. Milk production. Rural exodus. Technology. Innovation.

1. INTRODUCTION

In recent years, agriculture has suffered from the significant increase in the population in urban areas, showing a consequent decrease in rural areas. Despite the fact that industrial work in urban areas offered the employee great benefits in the form of a work permit, less manual labor and a fixed monthly salary, it didn't bring the desired professional fulfillment. Even so, the rural exodus was still persistent, until the arrival of a new scenario of technological innovation, which brought more agility in production, increased productive capacity that was performed exhaustively and late.

Milk production is of great importance to Santa Catarina's economy and its high productivity reaches the whole country. It is considered an activity carried out by the majority of family farmers in our region. Innovation and technology used to be seen more frequently in industry, but once they were introduced to the countryside, they became easier to use in rural areas, with equipment, machinery, fertilizers and inputs providing more and more support for farmers' activities. Dairy farming has become more technical over the years, from manual milking to mechanical milking, which requires more care, but is less costly and results in more milk being extracted than in the past.

Many of the farmers who used to produce culturally didn't adapt to the new procedures, and as a result there was a need to specialize in this production in order to minimize costs and increase milk productivity. The young people who wanted to succeed the family farm were more likely to learn and pass on to their parents, whose intention was not to remain in the business with the family. However, many of these young people with technical training, degrees, etc., due to a lack of opportunities in the city, or even unsuitability for industrial activity, return to the countryside to dedicate themselves and transfer their knowledge to family production.

In the face of innovation and technology in rural areas, the great participation of technological methods in milk production was demonstrated, along with academic technical training, which together have benefited agriculture and the production process. The prospect of family succession in dairy farming was noted. It is estimated that young people will return to the countryside, using the knowledge and technology they have acquired to encourage the development of agricultural production in the coming years.

1.1 RESEARCH PROBLEM

Over the years, the rural exodus has increased, resulting in fewer producers on farms. However, there is currently a reversal in this scenario, with many young people and adults with training returning to the countryside. With this in mind, this paper seeks to answer the following research problem: Are technology and knowledge contributing to the strengthening of family farming in milk production, together with the prospects for succession, in the municipalities of Atalanta, Agrolândia, Braço do Trombudo, Rio do Oeste and Trombudo Central, in the current period?

1.2 OBJECTIVES

1.2.1 General

- Analyze the prospects for family farming in the selected municipalities of the Alto Vale do Itajai in 2017.

1.2.2 Specifics

- To gather information based on unstructured interviews for the analysis of each municipality;
- Demonstrate the use of technological innovation and training to benefit milk production;
- Diagnosing the prospects for family succession on farms, as well as the return of the young rural population to agriculture.

1.3 BACKGROUND

The rural exodus has been a major challenge in recent years due to the migration from the countryside to the big cities, in search of a favorable economic condition that in agriculture was not sufficient or capable of meeting farmers' expectations. With the introduction of new farming machinery, it became possible to grow crops with higher yields and less manual labor. In the face of innovation and technology in this sector, many of the producers who migrated to urban areas are now returning to farming, showing a completely reversed scenario.

Through the use of technology, many producers seek training that would enable them to enter the urban environment. However, it is possible to bring this sector into the rural environment because it is innovating, causing farmers who used to produce in a cultural way to lose out on their production because they don't have the skills to support the machinery, even in their production process.

Through a field study, this paper seeks to demonstrate the benefits of technology and acquired knowledge in milk production, as well as whether there is a prospect of family succession and whether there is a possibility of young people returning to the countryside in the municipalities of Atalanta, Agrolândia, Braço do Trombudo, Rio do Oeste and Trombudo Central.

The data acquired in this research will serve as a basis for farmers to assess the young people's prospects of succession. It will help to analyze their training in the use of technology and innovation, so that they can apply it to their production, avoiding less costly productivity. For the academic world, this work will serve as support for future academic work, and can be applied as quantitative research, in which it will be possible to measure statistical data relating to the subject addressed, leading to more precise results.

2. LITERATURE REVIEW

This chapter will look at development in agriculture, with concepts of family farming together with the rural exodus, demonstrating the innovation and technology applied to the agricultural sector and the views of young people in the countryside in recent years.

2.1 AGRICULTURE AND DEVELOPMENT

According to Arbage (2006), among the main economic concepts linked to agriculture are economic growth and economic development. The distinction between the two is that economic growth emphasizes the expansion of economic indicators, while development seeks to quantify social indicators.

According to Souza (2011), rural areas had a higher concentration of population and the primary sector was the biggest generator of national wealth. As the economy developed, activities shifted to urban areas. Until the mid-1960s, agriculture was seen as a passive element that depended on economic stimuli from the public and industrial sectors. There was a certain discrimination against agriculture as a mobilizer of resources for industry. The Soviet industrialization of the 1920s and 1930s, which inspired this view, led to an intense flow of migrants, with resources going to industry as the poorest people moved from the countryside to the cities.

According to Souza (2011), this shift to the urban sector leads to the assertion that the implementation of industrial projects and sectors would certainly have a revolutionary impact on agriculture, resulting in an increase in the demand for food and raw materials and expanding the agricultural supply. The urban-industrial impact model thus considers agricultural development to be dependent on urban-industrial growth.

At the end of the 1960s, there were changes in this approach, and a balance between the urban and industrial sectors began to be advocated. In the following decades, the focus was on the influence of self-sufficient stimuli on the rest of the economy. The role of agriculture in the process of economic development was encouraged (SOUZA, 2011).

The agricultural sector would provide growth for the other sectors, which would be proportional, through the multiplier effect. As a result, there would be a structural change, with the share of the local product falling in the long term. As a result of this change, agriculture itself becomes responsible for the differentiation between agricultural and industrial productivity. Industrial progress would find its origins in agricultural support and economic development. As a result of this role played by agriculture, it ended up suffering as a result of the energy crisis, the increase in fertilizers and the fall in the prices of some agricultural products (SOUZA, 2011).

The link between agriculture and the rest of the economy derives from five basic functions, highlighted in Table 01.

Table 01 - Basic functions of agriculture

	Basic functions
1	To free up labor to be employed in industry and thus avoid raising the wages paid, so as not to depress the rate of profit and ensure the continued accumulation of capital;
2	Supply of food and raw materials for the urban-industrial sector, as demand grows with the development and intensification of the urbanization process;
3	Generate foreign currency by exporting agricultural products to finance development, purchase imports and repay foreign debt;
4	Transfer of savings for investment in industry and for the implementation of economic and social infrastructure;
5	Build markets for industrial goods, complementing urban markets;

Source: Based on Souza (2011)

According to Souza (2011), the basic functions of agriculture are described in table 01. The first function is the process of industrialization and urbanization of food products. When there is a smaller supply, the cost of living and wages rise, which contributes to a reduction in the rate of profit and the accumulation of capital. On the other hand, when the supply of industrialized food is greater, the production of industry and agriculture is spread out, thus favoring economic growth, as it generates a higher level of employment and income. The intense process of agro-industrialization achieved through the transformation of agricultural raw materials, and the use of fertilizers, equipment and machinery, fall under the second and fifth functions, as they are related. Within the five basic functions, growing agro-industrialization helps to add value to the product, making it more profitable for the raw materials. And rural capitalism, "[...]implying the more intense use of modern inputs and scientific management." (SOUZA, 2011, p.202). The use of inputs raises the cost of production and, as a result, reduces the producer's profit, which makes it impossible to inject a capitalist company, but it does make family production possible, where the income covers everything from the family's wages to its other costs.

In addition to the functions already mentioned according to Souza (2011), there are five other complementary functions that demonstrate government intervention as an incentive for agriculture to play a contributing role in economic development. Among the additional functions are: seeking a balance in food prices; reducing the rate of inflation; producing surpluses for export and increasing the production of capital; mobilizing resources for rural public investment; collaborating to reduce poverty and raise the productivity of labour and land; preserving the environment, reducing the emission of polluting gases which, as a consequence, will reduce the greenhouse effect.

Souza (2011, p.203) states that "government intervention in the formation of agricultural prices is justified when it increases the efficiency of the economy and improves income distribution and the well-being of the poorest population". A much lower price has a negative impact on the decision to plant, which increases prices in the long term, thus moving capital from agriculture to other sectors, generating inflation and reducing supply in the long term. The government must intervene competently in price formation. The most contributory policies are those that increase land capital, as they increase the product and generate an increase in the farmer's standard of living, as well as reducing the rural exodus (SOUZA, 2011).

For Souza (2011), the increase in productivity, which generates surpluses for export and domestic production, would mean that new agricultural areas would be expanded within the country. This certainly has negative effects on the environment, both because of the deforestation of resources and the use of fuels, herbicides and fertilizers, which are responsible for polluting water resources. Consumer society is also concerned about the use of chemical residues in food.

According to Azevedo and Pessôa (2011), modernization in the rural sector has brought about a bimodal system of family and employer agriculture. Family farming is of the utmost importance to Brazil, both in terms of generating income and jobs, but it is not highly valued by the state through public policies, compared to other sectors such as large-scale farming. In 1990, with the creation of the National Family Strengthening Program (PRONAF), the lack of resources was recognized and, through state investments, family farming, which until then had not been recognized, began to be covered.

According to de Azevedo and Pessôa (2011), Brazilian agriculture portrays complexity and specific conditions within the family farming and patronal farming models. It is possible to classify them according to their characterization, as they comprise an opposite differentiation in social terms, as shown in Table 02.

Table 02 - Characteristics of family farming and large-scale farming

Patronal Agriculture	**Family farming**
Total separation of management and work factors	Management and work closely related
Centralized organization	Production process run directly by the farmer
Emphasis on specialization	Emphasis on diversification
Emphasis on standard agricultural practices	Emphasis on the durability of natural resources and quality of life

Predominance of wage labor	Wage labor is only complementary
Technologies aimed at eliminating "field" and "momentary" decisions	Immediate decisions, suited to the high degree of unpredictability in the production process
Technologies mainly seek to reduce labor requirements	Decisions taken "on the spot", conditioned by the specificities of the production process
Emphasis on the use of purchased inputs	Emphasis on the use of internal inputs

Source: Azevedo and Pessôa (2011, p 485).

According to Azevedo and Pessôa (2011), the description of the characteristics mentioned in table 02 refers to the employer model, which used to be more favored, in contrast to family farming, which came later. The workforce in family farming is made up of the owners, who are family members, always seeking to use internal inputs. The opposite is true in patronal agriculture, which uses salaried labor and seeks external inputs to produce a given product.

2.1.1 Conceptions of family farming

According to Gasson and Errigton, (1993) apud Abramovay (1997), based on a few characteristics it is possible to define family farming, where the management is carried out by the residents who own the property, who form the workforce used with other family members, their assets and acquired assets come through the transfer of past generations. This definition demonstrates the specific income-generating capacity and size of these production units. It can also be seen that many family members do not live on the property, and the transfer of the business to non-family members is not very continuous.

According to Brasil (2006, p.?), under Law No. 11.326 of July 24, 2006, family farmers are those who meet the following requirements:

> I - Does not own, in any way, an area larger than 4 (four) fiscal modules;
>
> II - Use predominantly their own family's labor force in the economic activities of their establishment or enterprise;
>
> III - Have a minimum percentage of their family income originating from economic activities of their establishment or enterprise, as defined by the executive power; IV - Run their establishment or enterprise with their family.

According to Tinoco (2008), although family farming has a broad conceptualization, it is based on the workforce, the size of the property, the family itself carries out the activities and assumes the means of production, generating income from agricultural activity.

According to the FAO (Food and Agriculture Organization, 2014), family farming is considered to be the organization of various types of production, both agricultural and livestock, which are managed by the family itself and its workforce. Some of the factors favoring the development of family farming include territorial conditions, access to the market, credit lines, available natural resources, obtaining technology, education and extension.

2.1.2 Characterization of family farming

According to Abramovay (1998), family farming can be defined as having measurable value, as well as contributing to economic and social growth. The aid it receives is a result of its performance, as it contributes to sustainable development, generating better living conditions and, above all, reducing poverty. His vision associates a sector with the capacity to give a new emphasis to work in rural areas, which used to be seen as an activity that didn't compromise the construction of improved living conditions. "Many believe that family farming is just a new name for situations already known and characterized by expressions such as 'small producer', 'low-income farmer' or even 'subsistence units'." (ABRAMOVAY, 1998, p.145).

According to Schneider (2003), the term "family farming" only appeared in Brazil in the 1990s, and two events contributed to its existence. In the political field, there were demonstrations by rural unions which faced a number of challenges, including the opening up of trade, the unavailability of credit for farmers and falling prices for exported agricultural products. With the knowledge of family farming, settlers, tenants and partners were offered integration into the agro-industry. Another event was the government's 1996 loan for the creation of PRONAF (National Program for Strengthening Family Farming), which was the result of pressure from rural trade unionists, with the aim of promoting agricultural credit and supporting small rural producers, who had been excluded from public policies since the 1980s, which made it difficult for them to remain in these activities.

According to Vilkas and Nantes (2007), the management of family farmers is somewhat complex, since it depends on natural resources, seasonal markets, product conditions, animals and the weather, and it is completely impossible to make immediate changes to production, i.e. once you have invested in the production process, you can't abandon it, you just have to wait until it is ready to be cultivated. It is a complex management process for the producer, dealing with environmental, political, social and market aspects, as well as the resources coming from his company, which is the property.

According to Vilkas and Nantes (2007), in order to make the rural sector competitive, producers are improving the way they produce, with many using added value to their products as a way of differentiating them from other producers. Adding value opens up new markets for these producers, but they should be aware that this differentiation requires a risky investment compared to existing production.

According to Silva and Grossi (1998), in the face of modernization and technology in rural areas, agriculture has started to adopt pluriactivity in its activities in order to serve specific market sectors. As a new type of farming, there is also an increase in the value of rural areas, especially in activities such as leisure and housing, through fishing, ecological tourism, rural inns and farm hotels. This proliferation of pluriactivity has contributed to the creation of new jobs and a new means of income for property owners, with many industries moving to rural areas because of the proximity of raw materials and the reduction in labor costs.

2.2 RURAL EXODUS

According to Froehlich et al. (2011), the Brazilian population has declined over the last 50 years. The rural exodus was more frequent in the 1980s in relation to gender and age, and the conservative modernizations in agriculture contributed to the population moving towards urban areas, which also demanded cheap labour with the accelerated process of industrialization. Even though the reduction in the rural population is similar to what happened in Brazil, it cannot be compared with other European countries and the United States, because the big cities did not have enough control and preparation to serve the rural population. This resulted in a large concentration of slums in urban centers.

According to Politica Agricola (2009), after the 1960s, the rural population was occupied and grew between 1940 and 1985 and many people were employed in rural areas. However, in subsequent years, rural occupation fell, both in terms of occupation and population. The drop was reflected in high wages in urban centers. Mechanized agriculture replaced a large part of the work and was advantageous because of the indirect salary, but over the years, young people in particular became disinterested in rural employment, due to a lack of motivation and specialization, as well as government aid for the unemployed in urban areas, which was not present in the countryside.

According to Tomei and Lima (2015), the development model of patronage agriculture from the 1960s to the 1980s was accompanied by a rural exodus, as it suffered great inequality compared to the urban sector. It was only in the 1990s, following rural social events, that the government began to develop public policies aimed at agricultural development.

According to Abramovay (1998), the rural exodus in the 1990s fell slightly compared to previous decades. However, during the 1940s, more than a third of the rural population migrated to the big cities, while the exodus during the 1980s was more than 12 million people, reaching 5.6 million individuals.

According to Abramovay (1999), the rural exodus became an alternative to contribute to economic growth and also to the reduction of poverty in 1970. Families with a per capita income of less than

¼ of the minimum wage accounted for 43.9% of the total in the same year. In 1980, it showed a reduction of around 17.7%. It can be said that even with its precariousness, the urban sector in the 1970s contributed to improving the lives of the population moving from the countryside, bringing more opportunities that had been absent in the interior.

According to Abramovay (1999), the opposite happened in the 1980s, when the growth that was the precursor to the mobilization of the population was exhausted, with a slowdown in jobs in the most dynamic sectors, even though income generation did not mean job creation.

According to Abramovay (1999), even though Brazilian cities are very dynamic, they are subject to the arrival of migrants with precarious educational and professional backgrounds.

According to Tomei and Lima (2015), the consolidation of agriculture demonstrated the presence of sustainable development, as well as the emergence of agricultural and non-agricultural activities, acting as an incentive for rural families to settle in rural areas, in return minimizing the rural exodus at the end of the 1980s, which reduced urbanization disorder.

2.3 INNOVATION AND TECHNOLOGY IN RURAL AREAS

This chapter will provide information on innovation, its conceptualization and the technologies it applies to rural areas.

2.3.1 Conceptions of innovation

According to Schumpeter (1988) apud Dos Santos, Fazion and De Meroe (2011, p.3):

> Economic development arises from changes in economic life, a process that begins spontaneously, discontinuously, without impositions, with its own initiative, creating prerequisites for new developments. As such, the author emphasizes the importance of large companies as the central pillar of economic development, through what is known as creative accumulation and the accumulation of non-transferable knowledge in certain technological markets and, above all, the capacity for innovation.

According to Dos Santos, Fazion and De Meroe (2011), several studies have been carried out analyzing companies' capacity for innovation and commercial performance, with technology being the main factor in the use of available resources, producing them in a distinctive way and making them competitive. Technological innovation serves as an essential tool for the production process, whether it's introducing a new product to the market or even conquering a new space to increase production.

According to Dos Santos, Fazion and De Meroe (2011), Schumpeterian analysis shows that companies seek technological innovation in order to maximize profits. And by inserting it into the productive environment, competitiveness is achieved in relation to the competition, which will also lead to an increase in profits.

According to Tomei and Lima (2015), in order to achieve socio-economic transformation, they must use the technology available on their rural property, paying attention to their needs, inserting culture into the resources they already have so that this process of innovation results. The production of milk and its transformation into cheese is already an innovative process, because from the point of view of an industry it is a common process in which there is no innovation in production.

According to Maillot (1996, p. 256) apud Abramovay (2000), he states that:

> The company is not an isolated innovator. It is part of the environment that makes it act. The history of territories, their organization, their collective behaviour and the consensus that structures them are all major components of innovation. Innovative behavior is therefore not national, but depends on variables defined at local or regional level.

According to Abramovay (2000), innovation is a territorial element in the generation of new resources, with major repercussions for rural development. It refers to the intensification of a certain process in order to enhance the value of a given region.

2.3.2 Use of technology in agriculture

According to SENAR (2016), although technological advances in agriculture are not as visible as in other sectors, they are present in the processes of automation and mechanization in rural areas. It has been possible to achieve precision in agricultural activities where, through a set of specific information, a more correct and dosed use of both pesticides and reduced labor is obtained, directly impacting on cost reduction.

SENAR (2016), points out some examples that demonstrate this evolution in rural areas, which occurs mainly in the machinery used for cultivation, such as drones and remote-controlled aircraft that can use electronics and computers (UAVs) are able to scan and respond to the condition of the property and its information, such as crop health and land use, thus directly contributing to an agility in the production process that was previously carried out by several farmers, demanding more time in this activity. Through the use of images, it identifies the situation of the leaves, showing the degree of pests, which often cannot be identified with the naked eye, and provides the correct way to reverse this condition. The field has started to manage its own production with adjustable harvesters, irrigation stations with sensors according to the weather forecast and other sensors that measure the suitability of the land, informing it of the nutrients it lacks with an alert system.

Schmitz (2010) comments on Participatory Technology Development (PTD), where there is an interaction between external professionals and the rural population that covers innovative issues. Within the project, through certain starting points, the location of formal and scientific knowledge is distinguished, the consideration of priorities that differ between researchers and producers, the realization of the recognition of formal research institutions and others with superior capacity and their contribution to technological adaptations, for expansion and diversity in the conditions of less favored producers. In this way, the DPT encourages a third kind of knowledge among extension workers and farmers, as well as contributing to the process of evolution in rural areas.

2.4 YOUNG PEOPLE IN THE COUNTRYSIDE

According to Schneider (2008), there has been an expansion of new knowledge, where many institutions promote educational courses and projects for specialization, such as the Promotional Education Program of the State of Espirito Santo (MEPES), the Center for the Development of Rural Youth (CEDEJOR), located in Rio Grande do Sul, Paranà and Santa Catarina. After this journey, many young people join unions and associations or seek opportunities in the big cities, thanks to their added knowledge. There are also young people who remain on their farms with tobacco plantations or other productivity-related activities. Many of those who migrate to urban centers admit that they didn't meet their expectations when they moved.

According to Schneider (2008), this can affect the commitment to family farming as a result of this exodus of young people. Through hereditary succession, which is made up of three instruments: the transfer of goods and assets to successors, the preservation of the main activity carried out and the control of the business of the property, this does not become archaic.

For Schneider (2008), the rural environment began to absorb similarities to cities, with easier access to transport, education, the internet and communication, which had an impact on various changes in family life, increased schooling and labor relations.

According to Abramovay et al. (1998), it is unlikely that the investments necessary for its development, such as land and capital, will take place if there are no successors on the family farm.

According to Abramovay et al. (1998), even in view of the opportunities offered by big cities, many young people claim that they will not achieve their professional goal through the current succession process. In the world, only a third of business owners say they have doubts about the succession of their business. "[...] Succession problems tend to appear predominantly in establishments that present the greatest difficulties for their economic viability [...]". The concern is based more on the existence of successors than on the actual condition of the property.

Within selective migration, there is more interest in girls compared to boys, because working in the countryside brings different attributions, as women do not have the decision-making skills for certain activities (ABRAMOVAY et al, 1998).

2.5 PRODUCTION CHAINS

According to Prochnik (2002), production chains reflect the correlation between economic agents. For Prochnik (2002, p. 1), it can be defined as "[...] a set of consecutive stages through which the various inputs pass and are transformed and transferred [...]", thus incorporating other production chains.

According to Zanella and Leite (2016), the production chain began to be referred to as *agribusiness.* When it arrived in Brazil, it came to be called an industrial complex, referring to all the products that are offered to consumers, not just those produced on farms. A production chain is described by Zanella and Leite (2016) apud Farina and Zylbersztajn (1992, p. 190) as "[...] the succession of stages of transformation through which the raw material passes, constituting a unified space for the generation and appropriation of profit and accumulation [...]".

According to Silva (2005), production chains can be understood as "systems" which, through a conjunctive interaction within a production process, offer products or services to consumers. By understanding the production chain, it is possible to identify its potential, missing factors and its problematization according to each segment, comprising both its internal actions, which can be exemplified by the organization of farmers into cooperatives, who purchase and market inputs, storing and marketing *commodities,* for their own benefit or transforming them into raw materials. As for external factors, these include the impact of changes to or the creation of taxes on the product, the barriers to exporting these products, their classification and standards, and the final process, which involves demands on the end consumer.

2.5.1 Conceptions of the milk production chain

According to Viana and Ferras (2010), the development of the milk production chain in the

Brazil marked its beginnings with the crisis of 1929, through the substitution of imports and the expansion of the consumer market due to the accelerated process of urbanization. It was only in the subsequent years, between the 1950s and 1960s, that the country underwent a process of transformation, with the emergence of roads, the installation of the equipment industry, the emergence of different types of milk, together with innovations in packaging and the boost given to the industrial process by multinational companies.

It was only at the beginning of the 1990s that industrialization took hold, advancing and opening up the market, which would have a direct influence on this chain, making it more competitive and government intervention less swift, as it leaves the permanence of price formation resulting from the market laws of supply and demand for this sector. It also highlights the importance of increasing technology in agribusiness for the development and competitiveness of this production chain (VIANA E FERRAS, 2010).

Santos and Marcondes (2007) state the importance of milk production for the state of Santa Catarina in social and economic terms. In 2004, its production accounted for 6.3% of the entire Brazilian population, making it the sixth largest producer in the country.

According to Banco do Brasil (2010), Brazil is the world's sixth largest milk producer and is considered the most important production in Brazilian agriculture. According to Do Brasil (2010, p.15), milk production has "[...] 1.3 million milk producers and a production of 27.5 billion liters/year, generating R$ 64 billion/year and employing 4 million people [...]".

Banco do Brasil (2010), claims that the milk production system does not provide adequate profitability for the producer, as it is not considered specialized and has a vast diversity in its production process. Many producers of small rural properties own 90%, with small production and poor technology, since there is an estimate of only 2.3% who have specialized in acting as rural companies.

According to Banco do Brasil, although there are few in-depth figures on this production system, there was an increase of 44% between 1980 and 1993, with both productivity and the number of cows increasing by 20% and 80% respectively. This shows that even though milk production in Brazil is lower than it could be, it has increased considerably in recent years. It is notable that the amount of milk produced has decreased, compared to the total number of cows, as a result of the lack of maintenance in extractive systems and the absence of technological processes for the production of cows in non-productive periods.

According to Banco do Brasil (2010), our country has great natural advantages over other countries, as well as climatic conditions that contribute to the production of grains, which has an impact on the cost of dairy production, favoring investments in this dairy sector. It also demands low costs, providing the potential to increase milk productivity.

According to Banco do Brasil (2010), the milk production chain is made up of a number of links that pass through each production sector. These links include:

Chart 03 - Links in the production chain

Production inputs:	They include other companies that supply veterinary medicines, fertilizers, pesticides, electricity, genetic material, milking and refrigeration equipment; Packaging, machinery for the dairy industry, starter cultures, electricity and thermal energy, vehicles for collection,

	transportation and storage;
Primary production units	It is estimated that there are around 1.3 million farms in a chain that involves around 4 million people in Brazil;
Intermediate marketing units:	Milk producers' associations are active in the refrigeration or collection process, collectively, in order to market the milk to cooperatives and dairies, as well as collectively purchasing inputs and contracting technical assistance services, among others; Milk cooperatives make up around 300 to 350 active members, where 40% of the formal milk produced in Brazil is purchased;
Processing plants	Companies of different sizes, which are responsible for purchasing and processing around 60% of Brazil's formal milk production
Research, development, training and technical assistance institutions and companies	Embrapa, Emater, CONTAG, FETAG, SEBRAE, SENAR, CNA, Ministry of Agricultural Development (MDA), Ministry of Agriculture, Livestock and Supply (MAPA), BNDES, Banco do Brasil, trade unions, among others.
Final marketing units	Wholesale chains, supermarkets, food companies, bakeries, snack bars, bars and restaurants.

Source: Adapted from Banco do Brasil (2010, p. 18-19).

According to Banco do Brasil (2010), table 03 shows that the milk production chain begins with milk producers, production inputs and development institutions, and then interconnects between cooperatives, associations and private dairies, reaching supermarkets, bakeries and the places that distribute the products to consumers.

Investments in research, animal feed, genetic improvement, special equipment for milk refrigeration and milking, among other factors, have contributed to increasing the competitiveness of the milk production sector. This has also been seen in the supply of inputs, rural producers and even retail establishments (VIANA E FERRAS, 2010).

According to Banco do Brasil (2010), cattle farming is exposed to a number of challenges that restrict production in the milk production chain, encompassing geopolitical and social aspects of a given region. These include a) Poor planning and management of the activity; b) The unit where production takes place is not worked on in a systemic way or visible as an organization; c) Production and expected profitability are scarce; d) Lack of adaptation to the technological process; e) Government policy for the sector; f) The supply of the sector; g) Marketing channels; h) Genetic process of the herd; i) Systematization of the production chain; j) Regarding the price of the product; k) Targeted technical assistance; l) Promotion of access to markets; m) Lines of financing for the sector; n) Capacity building and training; o) Tax incentives; p) Inspection by the competent bodies and; q) Unsatisfactory monthly income.

The Bank of Brazil (2010, p.21) also points out that:

> Considering primary production as the weakest link in the milk production chain, where specific actions can generate positive transformations, with significant results in the overall context of the chain, this is the link that deserves the most attention and investment. There are ways to transform this reality. It's not a question of investing in research to generate knowledge, because there is already knowledge and technologies that can be passed on to producers, generated by state and federal government institutions such as Embrapa.

According to Banco do Brasil (2010), there are factors in the milk production sector that demonstrate the complexity and variables of the production system.

Table 04 - Factors in the milk production system

Herd structure	Number of cows, number of heifers % lactating cows, other categories;
Weather	Temperature, rain, frost;
Area	Topography, fertility, drainage, pasture, forage crops;
Management	Improvement, health, disposal, control, identifying problems, solving problems;
Man	Participation, commitment, knowledge, management skills;
Others	Energy, taxes, transport, etc;
Installations	Technical knowledge, suitability functionality, sizing requirements;
Market	Season, milk quality, animal sale, location, purchase of inputs, purchase of feed;
Machinery	Cost of labor, quality of labor, operation, use, replacement, maintenance, necessity;
Lactating cows	Quality, breeding, nutrition, comfort;

Source: Banco do Brasil (2010, p. 21).

According to Banco do Brasil (2010), there are three factors that contribute to competitiveness in this sector: the rational use of natural resources, technological innovation and modernization in the organizational management system, flexibility and market development.

2.6 TECHNOLOGY IN DAIRY FARMING

Achieving efficiency in livestock production is intrinsically linked to technology, which contributes to higher productivity and lower fixed costs. Rural companies that adhere to technological production have high productivity rates compared to those that maintain their production under traditional livestock conditions. Technology in the dairy industry helps with various aspects, such as herd management, better care and conditions for the herd, genetic improvement, attracting machinery and a skilled workforce, among others (BANCO DO BRASIL, 2010).

It is no longer the custom to remove the milk by hand and then fill the bucket. Nowadays, many farms already use mechanical milking to transport the milk from the animal to the cooler (DIARIO CATARINENSE, 2010).

According to Banco do Brasil (2010), some of the technological processes that have emerged have been extremely important for the proliferation of this activity, including the use of artificial insemination, which has led to improvements in herd genetics, even though its use is applied among large producers. It contributes to disease control, cross-breeding, accident prevention, herd standardization, accident prevention with producers, among other favourable conditions. Mechanical milking, which has gradually reduced labor costs, is recognized on medium and large farms. There are models with advanced technology that even demonstrate individual monitoring of production. Another technological contributing factor is the controlled health of dairy cattle, where they undergo a program of vaccination, deworming and the correct use of the main vaccines for the herd, against foot-and-mouth disease, rabies, paratyphoid, among others. The diet has undergone a great deal of development, with cultivated pastures and selected grasses.

According to Diàrio Catarinense (2010), even the cows' breasts were only washed before the milk removal process, but recently they have been washed with a special product to protect them from milking, thus avoiding the spread of diseases while the animal is in the field, as well as to improve the quality of the milk supplied to the dairies.

Banco do Brasil (2010) points out that developments have taken place in production techniques, such as rationed grazing, which is the alternation of pasture use, allowing the animal to use the pasture for a certain period, followed by a rest period which serves as a form of nutrition, producing fodder for grazing. In this way, in addition to preserving the environment and consequently increasing productivity, the occurrence of ticks is reduced, as their life cycle includes the cattle's rest period. Confinement, where lactating cows remain in a covered area, provided with food and milking, only occurs on a scale of high productivity. Feed control, the technological process that assesses the urea content in the milk through the feed, gives positive results in terms of the correct amount, resulting in less waste and an improvement in the animal's nutrition.

With regard to facilities, Banco do Brasil (2010) states that they are of the utmost importance as they facilitate animal handling, influencing productivity and health. It is involved in the choice of the facility to be used because it has a direct impact on costs, functionality and durability. Within the context of facilities for milk production are the milking parlor, the use of fences, stables, salt troughs, among others. The equipment to be used for production can be found on the website of the Brazilian Agricultural Research Corporation (EMBRAPA), from manufacturers of milk coolers and cleaning products to a list of tank manufacturers.

On most small farms, this is done by outsourcing a forage harvester, or ensilage machine, which can produce an average yield of 10 tons per hour. Maize silage is first planted by choosing the best space and soil condition, which should be fertile, flat, sloping slightly and very close to where the silage process will take place. With adequate fertilization, planting begins, which usually takes place between September and November, due to the amount of rainfall during this period. Shortly after planting, the harvest point is reached, when the plant already shows dry leaves and the straw of the cob is yellowish. If the corn is cut before the correct time, this has a direct impact on the quality of the silo (NOVINSKI, 2013).

In terms of dairy technology, around 35% of the total produced in Brazil is fluid milk, the remaining 33% is cheese production, 26% is powdered milk and 6% is other derivatives (BANCO DO BRASIL, 2010).

2.6.1 Milk control and quality

Among some of the diseases that affect cows, such as mastitis, it is of the utmost importance for farmers to take part in training courses on hygiene, the physiology of lactation, how milking equipment works and how to maintain it, in order to be able to deal with such situations (MÜLLER, 2002).

According to Silva (2016), the new estimates for somatic cell count (SCC) and total bacterial count (TBC) were due to come into force on July 1, 2015, in accordance with normative instructions from the Ministry of Agriculture and Supply (MAPA). However, it was extended for another two years due to its modification.

According to Granato and Nunes (2016) somatic cells can be understood as cells that pass through the cow's milk in response to some aggression in the mammary gland, a sanitary issue for milk evaluation. The fluctuation in the count of these cells depends on the increase in the concentration of bacterial infection. SCC varies according to age, stage of lactation, annual season and nutrition, stress and the main factor is mastitis. Mastitis can be understood as an inflammatory process in the mammary gland caused by physical, biological or chemical reactions. As a result of the inflammation, the vessels dilate and countless substances from the blood pass into the milk, leaving a lower milk secretion capacity.

The total bacterial count (TBC) is directly related to the hygiene, storage and transportation conditions of the milk. CBT and SCC work to prevent contaminated milk or milk unfit for consumption (SILVA, 2016).

For CBT analysis, tests are carried out to assess the microbiological quality of the milk, which can be a standard or total count of microorganisms. The result is obtained in number of bacteria, or also in colony-forming units (CFU) per milliliter of milk (mL), in order to carry out the necessary evaluations regarding the hygiene of production, as well as the storage of milk (EMBRAPA, 2017).

In short, the SCC is considered to be the animal's defense cells, thus fighting infection and repairing the mammary gland. If the count is high, it is an indication that the cow is going through a process of infection. CBT is

respectively the total number of bacteria present in 1ml of milk, so it is possible to measure the degree of hygiene used in the handling of milk (SANTA CLARA, 2014).

Table 05 - Quality parameters for refrigerated raw milk

rural property index	From 01/07/2008 to 31/12/2011	From 01/12/2012 to 30/06/2014	From 01/07/2014 to 30/06/2016	From 01/07/2016
CBT (CFU/mL)	750.000	600.000	300.000	100.000
SCC (SC/mL)	750.000	600.000	300.000	400.000

Source: Adapted from Granato and Nunes (2016, p. 52)

According to Granato and Nunes (2016), in cows from the second lactation onwards, as SCC doubles, they lose approximately 0.6 kg of milk per day, which is around 180 kg per lactation period. As well as reducing the amount of milk produced, this has an impact on the levels of milk components. In this way, milk with altered levels results in a product with an unsatisfactory composition.

3. METHODOLOGICAL PROCEDURES

This section will explain the methods and techniques used to carry out the research. Initially, the research is characterized as bibliographical and descriptive. Firstly, bibliographical research was used to draw up a research plan, which, through the portrayal of information and data, denotes the local reality (BONAT, 2009). For Lakatos and Marconi (2003), bibliographical research is an analysis of what has already been written on a given subject, where it is transcribed with new ideas and arrives at a certain recent conclusion. In this way, the work was based on bibliographical and documentary research which, through books, documents and articles already prepared, aligned the work with the field research.

In terms of modality, the research will be classified as inductive, where the aim is to reach conclusions with a certain broad content, which may be different from the initial premises. For Lakatos and Marconi, (2003, p. 86) "Induction is a mental process through which, starting from particular data, sufficiently verified, a general or universal truth is inferred, not contained in the parts examined". The inductive method takes place in three stages, namely the observation of phenomena, the relationship between them and finally, how they are classified. The work is based on the intention of finding out what the prospects are for agriculture, with a focus on milk production.

We used descriptive research, which does not aim to find solutions, but rather to describe phenomena. According to Bonat (2009, p.12) "Another important function of descriptive research is that it provides a broad diagnosis of the problem motivating the research". The approach of the work is characterized as qualitative, because it is about reaching certain possible interpretations of the phenomenon as a whole, since it is not limited to a count (BONAT, 2009).

We used field research techniques with unstructured interviews. According to Gil (2002, p.53) field research is "[...] developed through direct observation of the activities of the group studied and interviews with informants to capture their explanations and interpretations of what happens in the group".

According to Lakatos and Marconi (2003), the unstructured interview gives the interviewer the freedom to develop the questions in any direction that best suits him. The interviews were carried out by the researcher and the data collected was presented in tables, followed by explanations in text.

The research is also called census, because it does not measure the sample of a target public, but its total, conducting interviews with a family in each municipality (LAKATOS and MARCONI, 2003).

The municipalities of Atalanta, Agrolândia, Braço do Trombudo, Rio do Oeste and Trombudo central were initially selected with the help of the Santa Catarina Agricultural Research and Rural Extension Company (EPAGRI), which had already carried out research on them, because they have a much lower milk production compared to the amount they produced a few years ago, This is due to the lack of young people to take over dairy farming, together with the ageing of the current dairy farming population and the lack of technological investment, which has led to the possibility of dairy farming becoming extinct in these towns (EPAGRI, 2017).

4. CONTEXT OF THE MUNICIPALITIES

The municipalities of Atalanta, Agrolândia, Braço do Trombudo, Rio do Oeste and Trombudo Central are described below, including their location, history, economic and social indicators, as well as their milk production.

4.1 ATALANTA

The town is located in the Alto Vale do Itajai micro-region. According to the Brazilian Institute of Geography and Statistics - IBGE (2017a), the estimated population of the city is 3,254 inhabitants, in a total area of 94.192 km^2 and its population density is 35.03 people per km2. In the figure below you can see the municipality of Atalanta highlighted, along with its neighboring towns.

Figure 1 - Location of Atalanta

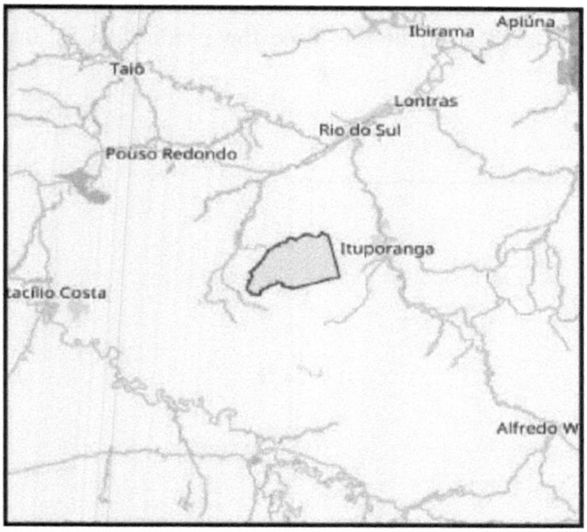

Source: IBGE cities, 2017a.

4.1.1 History

The municipality began to be colonized in 1930, due to the sale of allotments and land by Sociedade Colonizadora Catarinense S.A. It was known as Sede Dona Luiza, which later became known as Serra do Pitoco, because it is a mountain range that ends quickly. The Germans, Italians and Poles who colonized the region first cleared the forest to start farming and then built their homes. The first families to settle in Atalanta were located in Dona Luiza, Ribeirao Matilde and Alto Dona Luiza. They gradually settled in the town, and soon afterwards, with the opening of new roads, new families moved in to develop industry and commerce. Some time later, the region came to be known as Atalanta, because one of the directors of the Colonization Company, José Domingos Paglioli, named it after his grandparents who had come from Italy (IBGE CITIES, 2017b).

According to Atalanta (2017), the municipality of Atalanta has strong ties to agriculture, which has a direct influence on its economy. Mainly with regard to agri- and ecotourism, which have taken over much of the activity in the town, some farmers offer lodging, food and leisure on their property.

4.1.2 Economic indicators

The municipality of Atalanta is dominated by agriculture, where the main products are onions, tobacco, beans and corn. In the livestock sector, dairy farming stands out with the intention of commercialization. In the industrial sector, the city's main activities are dyeing, sawmills, factories, marble works, agricultural machinery, among others. Atalanta also sells food, furniture, agricultural products, bars and restaurants, cement products, among others (ATALANTA, 2017).

The municipality's main economic indicator is the Gross Domestic Product (GDP), the sum of all the services and finished goods produced in a given region, in this case the production of the municipality of Atalanta.

Graph 01 - Gross Domestic Product by sector in Atalanta

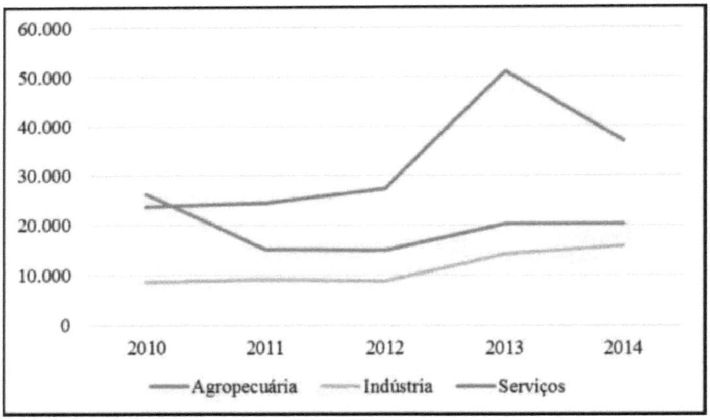

Source: Own elaboration using data from IBGE Cities, 2017a.

From Graph 01, it can be seen that even though the municipality has great roots in agriculture, between the years analyzed from 2011 to 2014 the services sector was the one that contributed to the city's economy, showing its growth especially in 2013. The agricultural sector followed, slightly ahead of industry, but at practically the same pace, rising in 2013 and remaining constant.

4.1.3 Social indicators

Below are the social indicators for the municipality of Atalanta that contributed to the analysis. These include the HDI, life expectancy, rural and urban population and the Gini index.

It is possible to understand that the "Human Development Index (HDI) is a summary measure of long-term progress in three basic dimensions of human development: income, education and health" (UNDP, 2017).

The Municipal Human Development Index (MHDI) measures the social indicators of a given municipality, state or country. The MHDI ranges from 0 to 1, and the closer it is to 1, the better it is considered to be. In the municipality of Atalanta, the index was 0.476 in 1991 (HDI 0.499) and is classified as very low. In 2000, Atalanta had an index of 0.599, which is low (HDI between 0.500 and 0.599). In 2010, the index was 0.733, classified as high (HDI between 0.700 and 0.799). Through the indices measured, the municipality has shown growth in the social sphere (ATLAS BRASIL, 2017).

Graph 02 below shows that the rural population of the municipality of Atalanta has been decreasing over the years, while the urban population has been increasing. In 1991, 2,722 inhabitants lived in rural areas, while 980 lived in urban areas, showing a great disparity between the populations, with the majority concentrated in rural areas. In 2010, the rural population fell considerably to 1,932 inhabitants, while the urban population increased to 1,368, where they have become increasingly close to each other, although there is a predominance of the rural population (ATLAS BRASIL).

Graph 02 - Rural and Urban Population of Atalanta

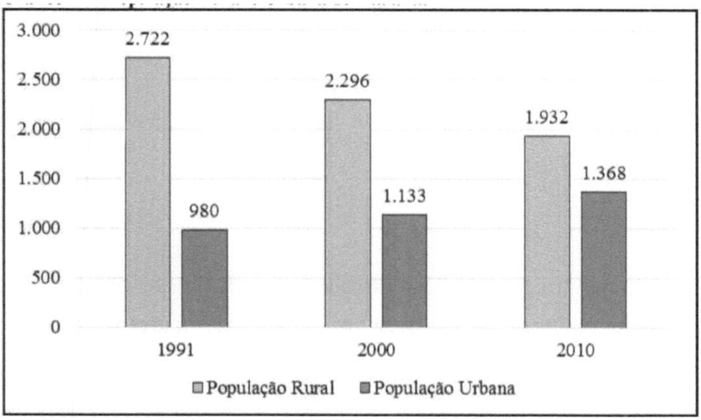

Source: Own elaboration with data taken from ATLAS BRASIL, 2017.

The life expectancy of a municipality is a very important index, as it shows the life expectancy of the population after birth. In Atalanta, the index in 2000 was 70.52 years, rising to 73.10 in 2010 (ATLAS BRASIL, 2017).

The Gini index is an indicator that measures the concentration of income. It shows the differences between the earnings of the richest and the poorest. It varies between zero and one, zero being total equality in income and one indicating total inequality, so the closer the index is to zero, the better the distribution of income (ATLAS BRASIL, 2017).

In the municipality of Atalanta, the Gini index in 2000 showed a coefficient of 0.50, rising to 0.44 in 2010 (ATLAS BRASIL, 2017).

4. 2AGROLAND

Agrolândia is a small municipality located in the Alto Vale do Itajai. According to IBGE (2017c), its estimated population is 10,581 inhabitants, in a total area of 207.554 km^2 and 44.92 people per km2. The figure below shows the location of the municipality that is highlighted, along with the nearby towns.

Figure 02 - Location of Agrolândia

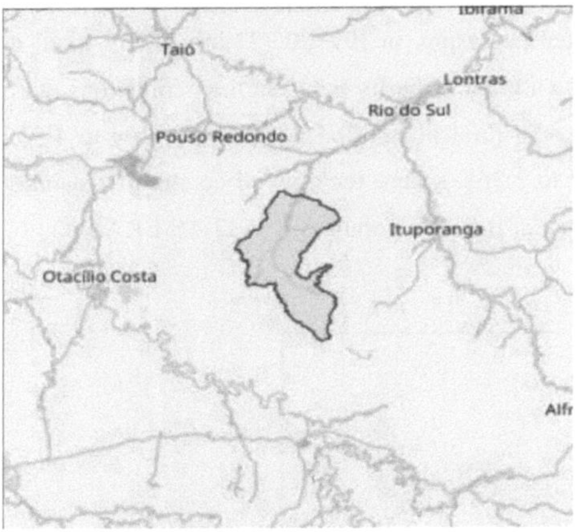

Source: IBGE cities, 2017c.

4.2.1 History

Two brothers from the municipality of Lages, Ricardo and Domingos Alves Paes, discovered the region by clearing it and starting to plant corn and beans, where they later settled permanently. German settler families moved in, and the region was known as Trombudo Alto (High Trombudo), due to the large curves and hills. Later, with the arrival of new Italian families from the surrounding towns, they dedicated themselves to agriculture and livestock farming, which are the activities that have developed the municipality to this day. In 1957, Trombudo Alto became part of the municipality of Trombudo Central until 1962, when it was dismembered and named Agrolândia, a name given because it was an agricultural town (IBGE, 2017d).

Agrolândia has great attractions in terms of waterfalls, mountains and rural tourism, and also strives to maintain the colonizing culture of the municipality (PREFEITURA AGROLÂNDIA, 2017).

4.2.2 Economic indicators

The municipality of Agrolândia has always been known for being a city that was at the forefront of agricultural activity. However, according to graph 03 below, which shows the municipality's GDP by sector, we can see how much industry has contributed to Agrolândia's GDP, and its considerable increase in recent years. The services sector, the second largest in the municipality, showed a small drop in 2013 along with industry, while agriculture increased, although in the last year analyzed agriculture was below the other activities.

Graph 03 - Gross Domestic Product by Sector in Agrolandia

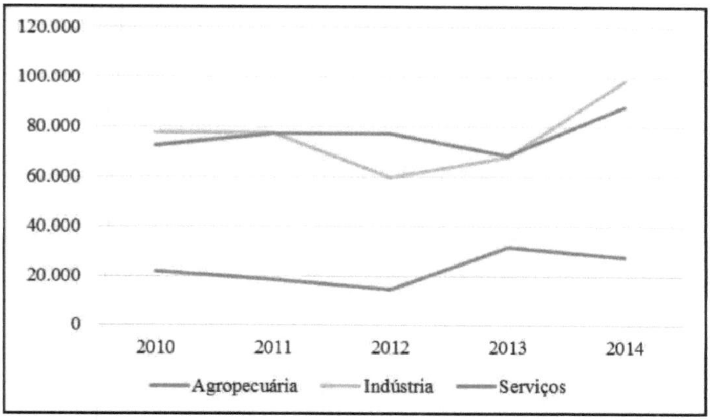

Source: own elaboration using data from IBGE cities, 2017c.

4.2.3 Social Indicators

Among the main indicators to be mentioned is the HDI, which in 1991 had an index of 0.458, classified as very low. In 2000, the index was 0.593, which is considered low. In 2010, the municipality's HDI increased to a relatively high 0.725 (ATLAS BRASIL, 2017).

The rural population in the municipality of Agrolândia, as shown in graph 04 below, fell from 1991 to 2000. However, there was a considerable increase in the population living in urban centers, which in 2010 was highly concentrated in urban areas, with 5,959 people. The rural population showed a slight increase from 2000 to 2010, but it is worth noting that the urban population has grown in recent years in the respective municipality (ATLAS BRASIL, 2017).

Graph 04 - Rural and Urban Population of Agrolândia

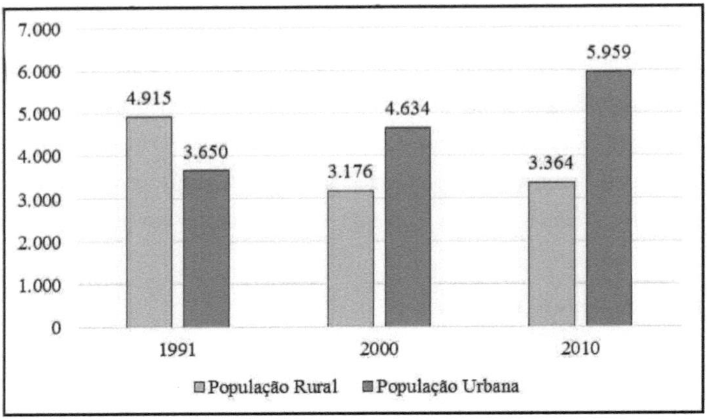

Source: own elaboration using data from ATLAS BRASIL, 2017.

In the municipality of Agrolândia, life expectancy in 2000 was 70.59 years, rising to 74.63 years in 2010 (ATLAS BRASIL, 2017).

The municipality's Gini index was 0.46 in 2000, and in 2010 it showed a good index of 0.36 (ATLAS BRASIL, 2017).

4. 3 TRUMPET ARM

The municipality of Braço do Trombudo, a micro-region, has an estimated population of 3,709 inhabitants, according to IBGE (2017e), in a total area of 90.319 km^2 and a population density of 50.59 people per km2. Figure 03 below shows the municipality in focus and the surrounding municipalities.

Figure 03 - Location of Braço do Trombudo

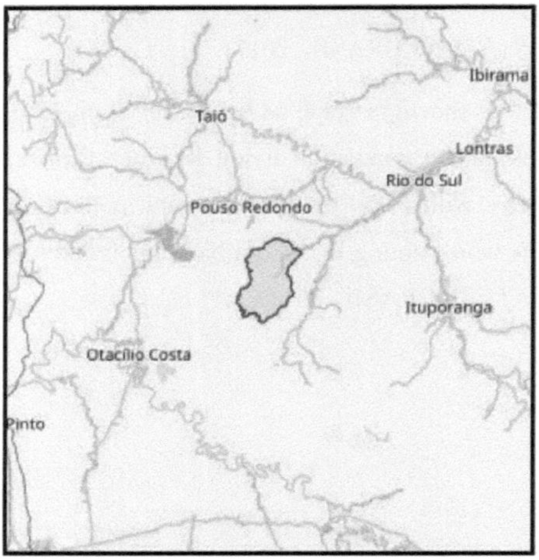

Source: IBGE cities, 2017e.

4.3.1 History

The town received its pioneers who came from the city of Blumenau and other neighboring regions in the mid-1920s and 1921s, where German families traveled through forests and along trails. At the time, Braço do Trombudo, before it was named, was known for being the gateway from the highlands to the coast of Santa Catarina, where due to the precariousness of these passages, journeys were first made on the backs of animals, later by carts until the arrival of trucks (PREFEITURA DE BRAÇO DO TROMBUDO, 2017).

The settlers of the time gave names to lands that were directly connected to nature, and for this reason, Trombudo became known because there were many tapirs in the region, the trombas being similar to the riverbed, which when they descended formed huge trombas. This gave rise to names such as Barra do Trombudo, Trombudo Central and Braço do Trombudo. It was only in January 1993 that Braço do Trombudo was emancipated and separated from Trombudo Central, to which it had previously belonged (PREFEITURA DE BRAÇO DO TROMBUDO, 2017).

Braço do Trombudo has natural beauties, such as the sulforosa waterfall, one of the best known, among other waterfalls and the Santo Antônio Chapel and Hill, which are tourist attractions in the municipality (PREFEITURA DE BRAÇO DO TROMBUDO, 2017).

4.3.2 Economic indicators

According to Graph 05, the municipality of Braço do Trombudo shows GDP by sector, indicating that in recent years industry has achieved high results for its sum, followed by services, which fell slightly in 2013 and increased again in 2014. And agricultural production has fallen and remained constant in recent years (IBGE CITIES, 2017e).

Graph 05 - Gross Domestic Product by sector in Braço do Trombudo

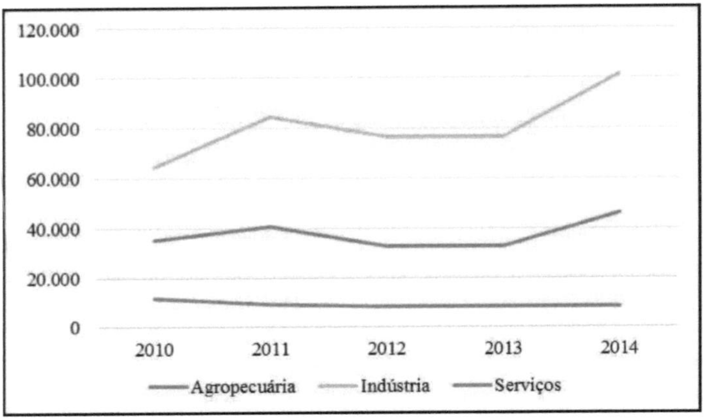

Source: own elaboration using data from IBGE Cities, 2017e.

4.3.3 Social Indicators

According to Atlas Brasil (2017), in 1991 the HDI of the municipality of Braço do Trombudo was 0.474, classified as very low, in 2000 it rose to 0.625, characterizing it as medium, and in 2010 it increased to 0.780, considered a high index.

Graph 06 below shows the rural and urban population of the town of Braço do Trombudo. According to the data, over the years the municipality has seen an increase in the population living in urban areas. However, the rural population also grew in 2000 to 1,565 residents, with only a slight drop in 2010 to 1,559 rural inhabitants. On the other hand, there were 1,898 urban residents in 2010. It is worth noting that in 2010 it is not necessarily the case that the population moved from the rural to the urban area, as it decreased to a much lesser extent, which may indicate that the population of other municipalities migrated to the urban center of the city (ATLAS BRASIL, 2017).

Graph 06 - Rural and Urban Population of Braço do Trombudo

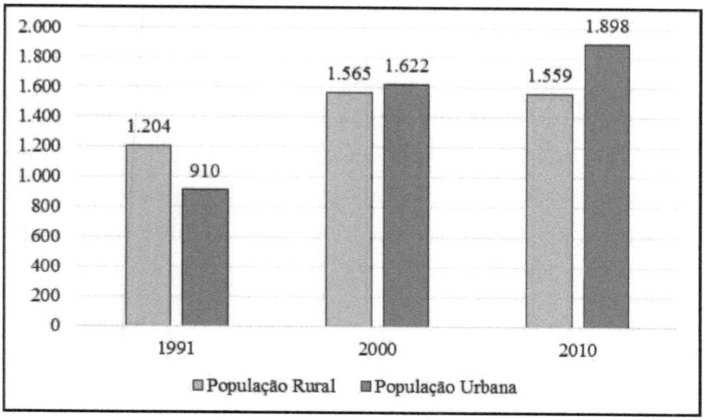

Source: own elaboration using data from ATLAS BRASIL, 2017.

According to Atlas Brasil (2017), life expectancy in Braço do Trombudo was 74.62 years in 2000, rising to 77.99 years in 2010.

Another social indicator is the Gini index, which in the municipality analyzed was 0.43 in 2000, rising to 0.36 in 2010 (ATLAS BRASIL, 2017).

4.4 RIO DO OESTE

According to IBGE Cities (2017f), the municipality of Rio do Oeste has an estimated population of 7,463 inhabitants, living in a territorial area of 247.805 km^2 , with a population density of 28.61 km^2 per person. Figure 04 highlights where Rio do Oeste is located, as well as the neighboring towns.

Figure 04 - Location of Rio do Oeste

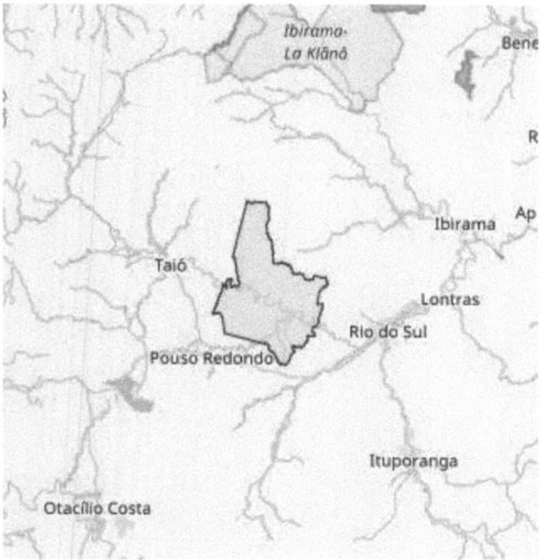

Source: IBGE cities, 2017f.

4.4.1 History

Rio do Oeste is a region that was colonized by Italians who came from Trento in 1912, through the suggestion of Friar Lucinio to some immigrants to found new communities near the banks of the Itajai do Oeste River, which due to the concentration of birds in the place, they named Barra das Pombas, caused by the junction of the Rio das Pombas with the Rio do Oeste (IBGE cities, 2017g; MUNICIPALITY OF RIO DO OESTE, 2017).

Towards the end of the 1920s, the name Barra das Pombas was changed to Vila Adolfo Konder, named after the visit of the governor who was thus named in 1927. However, the region was renamed Rio do Oeste after the Industrial Revolution in 1930. Dismembered from Rio do Sul, the municipality of Rio do Oeste was officially established on July 23, 1958 (IBGE Cities, 2017g).

According to Rio do Oeste (2017), the town's main attraction is the well-known State Polenta Festival, where around a ton of polenta is served. The municipality maintains ties with the Italian settler culture, in addition to its cuisine, with traditional parish festivals.

4.4.2 Economic indicators

Among the main activities carried out in the municipality of Rio do Oeste are farming, producing irrigated rice, tobacco, beans and cattle, among others. There is also clothing and ceramics manufacturing (PREFEITURA DE RIO DO OESTE, 2017).

Graph 07 below shows the municipality's GDP by sector, indicating that in recent years the services sector has contributed to Rio do Oeste's economic activity. It suffered a slight drop in 2013, when agriculture stood out until the following year; however, services continued to be the municipality's main activity in 2014, followed by agriculture (IBGE CITIES, 2017f).

Graph 07 - Gross Domestic Product by Sector in Rio do Oeste

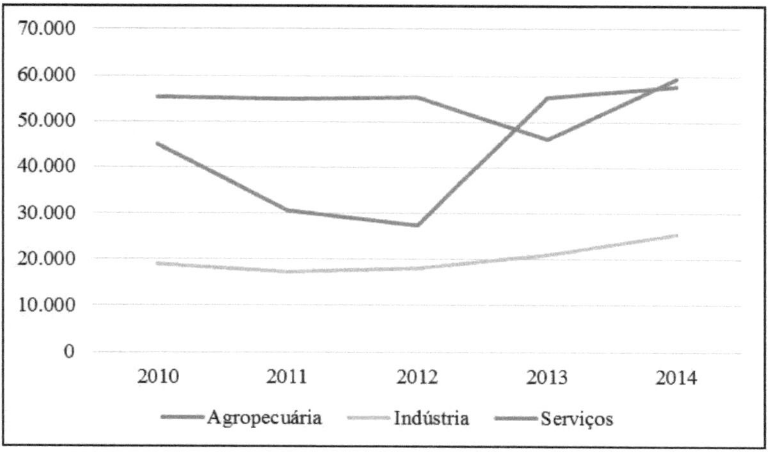

Source: own elaboration using data from IBGE cities, 2017.

4.4.3 Social indicators

The city of Rio do Oeste had an HDI of 0.490 in 1991, an indicator considered very low, rising in 2000 to 0.637, classified as medium, and in the last year of analysis, 2010, it had a high indicator, rising to 0.754, thus showing social development over the years (ATLAS BRASIL, 2017).

Graph 08 below shows the rural and urban population of the municipality of Rio do Oeste, with an increase in the urban population, although the municipality has always had a predominance of rural residents. In 2000, a total of 4,104 people lived in the countryside, compared to 2,626 in the urban center. In the last year analyzed, 2010, there was a considerable increase in the urban population to 3,390, but the rural population prevails, with 3,700 inhabitants in the same year (ATLAS BRASIL, 2017).

Graph 08 - Rural and Urban Population of Rio do Oeste

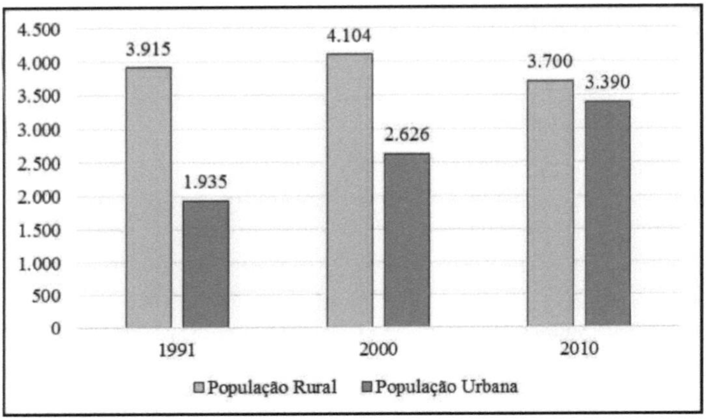

Source: own elaboration based on data extracted from ATLAS BRASIL, 2017.

According to Atlas Brasil (2017), the municipality of Rio do Oeste had a life expectancy of 74.23 years in 2000, rising to 78.52 in 2010.

The Gini index for the city was 0.44 in 2000, with a slight variation in 2010, showing an index of 0.47 (ATLAS BRASIL, 2017).

4.5 . CENTRAL TROMBUDO

The municipality of Trombudo Central is a small town in the Upper Itajai Valley. According to IBGE data (2017h), the municipality has an estimated population of 7,211 inhabitants, in a total area of 108.617 km^2 , with a population density of 60.33 people per km2. Figure 05 below shows the town of Trombudo Central and other neighboring municipalities.

Figure 05 - Location of Trombudo Central

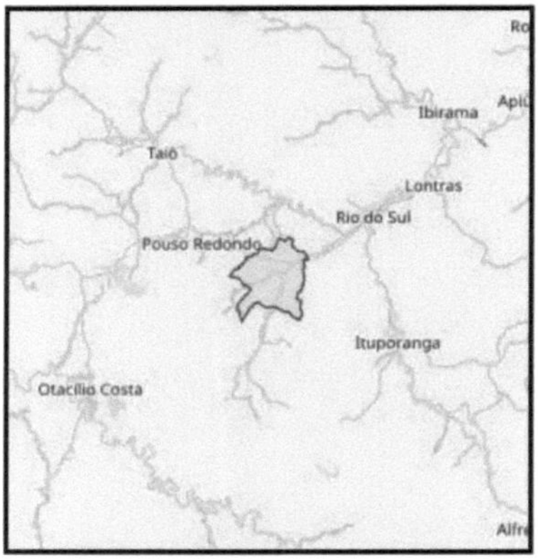

Source: IBGE cities, 2017h.

4.5.1 History

The municipality of Trombudo Central was initially colonized by German settlers. Other settlers came from the cities of Blumenau and Rio do Sul and explored the municipality in search of land for farming and livestock. The work they did was increasingly recognized in the municipality, which in 1957 was recognized by the governors of Rio do Sul, and the municipality of Trombudo Central was established on July 22, 1958. Its name has had many meanings, but the most common is that in the summer it was common for it to rain in the mountains where the rivers have their headwaters, and in the course of the rain it would cause an hour-long torrent of water or "tromba da água". Among other versions that gave rise to the name was the number of tapirs that roamed these lands (IBGE cities, 2017i).

The city bears traces of its colonizing culture, especially in the houses and structures of the municipality, such as the bandstand in the central square of Trombudo Central (PREFEITURA MUNICIPAL DE TROMBUDO CENTRAL, 2017).

4.5.2 Economic indicators

According to Trombudo Central (2017), Trombudo Central's economy is dominated by the primary sector, which extracts slate. It also manufactures screws, nuts, industrial furnaces and other activities linked to the tertiary sector. The municipality's main agricultural crop is manioc, but it also grows corn, tobacco, onions and beans, among others.

According to Graph 09, Trombudo Central's industry has been a major contributor to the municipality's GDP in recent years, only showing a decrease in 2012, while in other years it has been growing. The services sector has also stood out in recent years, being the second largest sector in the municipality's GDP (IBGE cities, 2017h).

Graph 09 - Gross Domestic Product by Sector in Trombudo Central

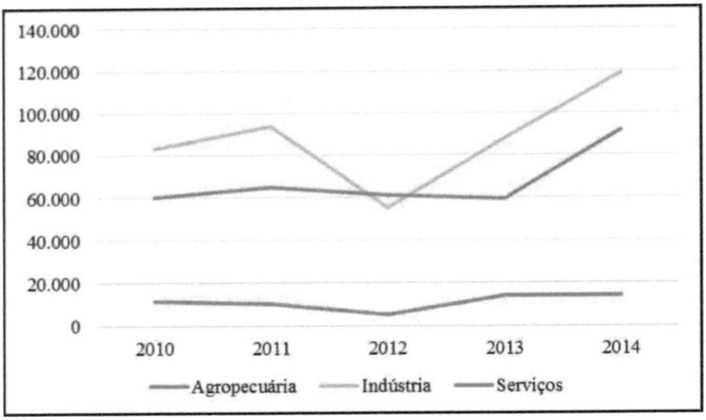

Source: own elaboration based on data from IBGE cities, 2017h.

4.5.3 Social indicators

The city of Trombudo Central had an HDI of 0.521 in 1991, which is considered low. In 2000, the indicator rose to 0.657, classified as medium. In 2010, the indicator was 0.775, characterized as high, showing a history of development in the social sphere (ATLAS BRASIL, 2017).

In Trombudo Central, according to graph 10 below, the concentration of people living in the urban area has increased in recent years. In 2000, 2,641 people lived in rural areas, while 3,154 lived in the urban center. This rural population fell to 2,452, while the urban population grew to 4,101 in 2010 (ATLAS BRASIL, 2017).

Graph 10 - Rural and Urban Population of Trombudo Central

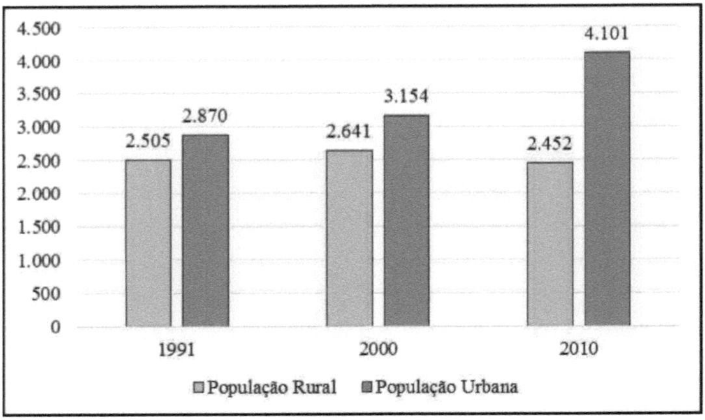

Source: own elaboration based on data extracted from ATLAS BRASIL, 2017.

According to Atlas Brasil (2017) the life expectancy index for Trombudo Central in 1991 was 71.82

years, rising to 73.80 years in 2000, and increasing in 2010 to 77.77 years of life expectancy at birth.

According to Atlas Brasil (2017), the municipality had a Gini index of 0.49 in 2000, showing a slight improvement in 2010 when the indicator fell to 0.44.

4.6 MILK PRODUCTION IN THE MUNICIPALITIES

Table 01 below shows milk production in the municipalities analyzed from 2005 to 2015.

Table 01 - Milk production in the municipalities from 2005 to 2015

	Agrolandia	Atalanta	Trombudo Arm	Trombudo Central	Rio do Oeste
2005	4.550	2.340	3.560	3.500	10.200
2006	4.550	6.500	3.560	3.325	11.560
2007	5.265	4.640	3.800	3.300	12.750
2008	6.080	3.800	5.790	3.420	15.700
2009	6.475	3.800	5.880	3.440	15.700
2010	6.475	3.420	5.680	3.440	12.000
2011	6.700	3.990	5.300	3.705	12.700
2012	7.193	3.640	4.950	3.990	14.400
2013	7.740	5.200	4.600	4.200	17.850
2014	9.120	5.200	4.600	4.200	16.779
2015	9.480	5.040	4.500	4.410	17.010
Total	73.628	47.570	52.220	40.930	156.649

Source: own elaboration using data from IBGE cities, 2017.

In the municipality of Agrolândia, milk production increased consecutively in all the years analyzed, from 7,740 liters of milk in 2013 to 9,120 liters in 2014. In Atalanta, production increased from 2,340 to 6,500 liters in 2006, but in subsequent years there was a slowdown in production, which only remained stable in 2013. In the municipality of Braço do Trombudo, production rose in 2008 from 3,800 liters to 5,790 liters, but over the years there has been a year-on-year reduction in production. In Trombudo Central, on the other hand, although there were some disparities between the years, it showed an increase over the years, reaching 4,410 liters in 2015. Graph 11 shows the milk production of the municipalities under analysis (IBGE cities, 2017).

Graph 11 - Milk production in the municipalities from 2005 to 2015

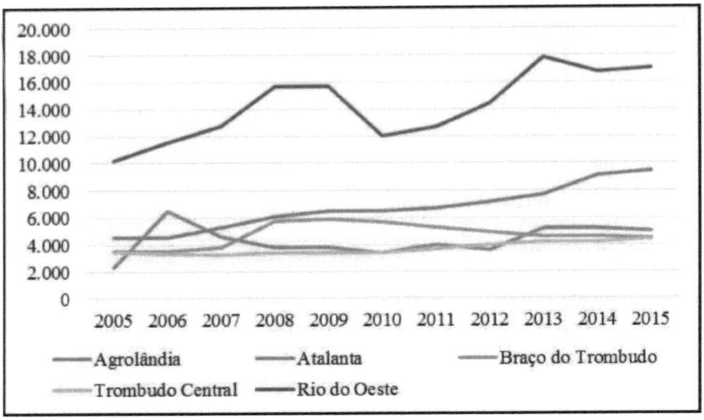

Source: own elaboration using data from IBGE cities, 2017.

The municipality that stands out with the highest production is Rio do Oeste, which had a high production in the period shown, with the highest amount of milk produced in 2013 reaching 17,850 liters, remaining only slightly below this figure in subsequent years (IBGE CITIES, 2017).

4.7 PROPERTIES

This section will explain all the properties where the interviews were carried out, as well as information, figures and tables describing their characteristics and data.

4.7.1 Marcelo property

Located on the Rio Caçador General Road, about 6 km from the center of the city of Atalanta. The property covers 36 hectares, where they graze dairy cows and grow tobacco. The family is made up of Genésio Valmor Marcelo (56 years old), his wife Maria Bernardino Marcelo (51 years old) and their son Jeferson Marcelo (23 years old), as shown in figure 06 below, with Genésio and his son doing the work on the property (MARCELO, 2017).

Figure 06 - Maria, Genésio and Jeferson Marcelo

Source: Author's collection (2017).

According to Marcelo (2017), the family started dairy farming as a succession, i.e. his parents were already involved in production and passed it on to him as the years went by. Table 05 below summarizes the characteristics of Marcelo's property.

Table 06 - Marcelo property data

Milk Purchasing Company	**N° Cows**	**No. of milkings**	**Approximate daily production**
Riolat	21	4	250 liters

Source: Prepared by the author based on the interview (2017).

The family *has been dairy farming for* more than 20 years and, as it was inherited from their parents, they intend to continue with it. Dairy farming presents some difficulties, such as monthly price fluctuations; at the moment they receive R$1.09 a liter for their milk. In 2013 there was a large production of milk throughout Santa Catarina, making milk production highly sustainable, where the price of milk fell to R$0.92 (MARCELO, 2017).

Since 2015, due to market concerns and the stricter standards for Somatic Cell Count (SCC) and Total Bacterial Count (TBC), many producers have stopped farming. The SCC is difficult to keep within the standards, as it is found internally in the animal and takes about a year to adapt to the established standards. CBT, on the other hand, can be brought up to the desired level within a month. These control analyses are carried out by the dairy itself, which buys the milk from the farm. Although milk production has some setbacks, it is still attractive for the family (MARCELO,

2017).

For their monthly income, the family depends on milk production, but also on tobacco production, which they receive according to the harvest, since the couple is not yet retired (MARCELO, 2017).

Figure 07 - Marcelo family property

Source: Author's collection (2017).

According to Marcelo (2017), his son will be the successor to the property and the dairy business. He says that the courses he has taken, including tractor and agricultural implement training - operation and maintenance through the National Rural Learning Service (SENAR); leadership, management and entrepreneurship with rural youth at the Research Company

Agropecuària e Extensao Rural de Santa Catarina (EPAGRI), subsidized by the government; and also participated in the Rural Entrepreneurs Program (SENAR), which contributed to his development on the property, in terms of keeping up with technological advances.

Figure 08 - Certificate of course taken

SENAR
Santa Catarina

CERTIFICADO

O Serviço Nacional de Aprendizagem Rural - SENAR confere o presente Certificado ao (à) Sr. (a)

Jeferson Marcello

Identidade nº 000.054.56647 SSP/SC, por seu aproveitamento no (a) Treinamento em Tratores e Implementos Agrícolas - Operação e Manutenção - CBO 6410-15, realizado em ATALANTA - SC, no período de 20/10/2015 a 22/10/2015 com carga horária de 24 horas.

Florianópolis, 01 de dezembro de 2015

Source: Author's collection (2017).

The study acquired is increasingly necessary, as it brings knowledge to animal welfare, pasture technologies, especially in obtaining higher production at a reduced cost (MARCELO, 2017).

The Marcelo family has not adapted to many technological methods that the market has brought in recent years, only the basics to facilitate handling and productivity. In addition to mechanical milking, picketing the pastures for the cattle is considered one of these methods, as it makes the cattle walk only on a certain piece of land, reducing their journey and avoiding the loss of nutrients if they were to graze the entire property (MARCELO, 2017).

Figure 09 - Dairy cattle owned by Marcelo

Source: Author's collection (2017)

The rural youth population will grow over the years, as agriculture is becoming a profitable area. There is a study stating that in the next 20 years Brazil will be the world's largest agricultural producer, and this possible growth of the population in agriculture may occur (MARCELO, 2017).

4.7.2 Will Property

Located in the municipality of Agrolândia, about 3.5 km from the town center. The Will property covers 110 hectares, where the majority of the land is used for milk production, as well as pig production. Cristiano Will (39) is the owner, who together with five other employees is responsible for the dairy production (WILL, 2017).

Figure 10 - Cristiano Will

Source: Author's collection (2017).

Cristiano started dairy farming with the encouragement of his father, who had *been* dairy farming for over 35 years. The property has 300 cows, but only 150 produce milk at the moment. Table 07 shows some information about the property (WILL, 2017).

Table 07 - Will property data

Milk Purchasing Company	**N° Cows**	**No. of milkings**	**Approximate daily production**
Tyrol	150	6	3400 liters

Source: Prepared by the author based on the interview (2017).

Even though he has been working for some time at a multinational in the sales area, Cristiano has opted for dairy farming, saying that he has no intention of leaving this activity, but rather of investing in increasing production. Even though there are certain obstacles for dairy farmers, he believes that they must adapt to production, just like other agricultural activities. Currently, Cristiano sells his milk to the dairy at R$1.10 a liter. At the moment, Cristiano's dairy farming is a non-family activity, i.e. only with employees whom he has hired (CRISTIANO WILL, 2017).

Figure 11 - Will family property

Source: Author's collection (2017).

For Will (2017), the issue of price changes interferes, but in a reasonable way, since he has a high monthly production. As for the issue of bacterial control in milk, he says that it is a general problem for producers, and that they must adopt technology to alleviate this problem, and that it is also affected by the climate: when there is rain, the animal's grazing becomes more conducive to bacteria. In your case, some cows are suffering from mastitis, an inflammation of the mammary gland. The problem has already been identified in a vacuum system that has been idle for too long, so the producer should replace it so that it doesn't cause any more damage in the future.

Figure 12 - Dairy cattle owned by Will

Source: Author's collection (2017).

Due to the fact that Will's family has been involved in milk production for a long time, Cristiano attended a technical course in agriculture at the Vidal Ramos Professional Education Center in Canoinhas - SC, which gave him a broader view compared to other farmers. However, he didn't complete the compulsory internship, which prevented him from receiving his certificate of completion (WILL, 2017).

According to Will (2017), the succession process is still uncertain, as the owner and his wife have a small daughter, who has not yet decided whether or not to remain in the business.

For Will (2017), the study only gave him a basic idea of what farming is like. Like other activities, milk production requires knowledge and, above all, daily practice in order to evolve.

Figure 13 - Declaration of participation in a technical course

ESTADO DE SANTA CATARINA
SECRETARIA DE ESTADO DA EDUCAÇÃO
CENTRO DE EDUCAÇÃO PROFISSIONAL "VIDAL RAMOS"
CURSO TÉCNICO EM AGROPECUÁRIA – 2º GRAU
CAIXA POSTAL 381 FONES: (047) 3622-4050 (047) 3622-4324
89460-000 CANOINHAS – SANTA CATARINA

DECLARAÇÃO

Declaramos para os devidos fins, que Cristiano Will, portador do RG 3.458.258, nascido em 23/04/1978, filho de João Will e Marlene Will, frequentou em nossa Unidade Escolar o Curso Técnico em Agropecuária, nos anos de 1994 (1ª Série), 1995 (2ª Série) e 1996 (3ª Série), não concluindo o curso, pois não realizou o estágio supervisionado obrigatório para conclusão do mesmo.

Por ser verdade firmamos a presente.

Centro de Educação Profissional "Vidal Ramos".
Canoinhas (SC), 10 de novembro de 2017.

Source: Author's collection (2017).

With regard to technology, the owner says that there has been a big reduction in labor in the sector, as well as the ease of genetic production, since his property uses insemination in the cows for breeding. Cattle are also kept locked up to optimize production (WILL, 2017).

On the Will property, the technological methods used in dairy farming are artificial insemination, silo made from forage, mechanized milking, an electronic milk system, which transfers milk directly to the refrigerator, and a composting system, which provides more nutrients for the dairy cattle (WILL, 2017).

Figure 14 - Mechanized milking and electronic milk collection system

Source: Author's collection (2017).

A lot more young people are returning to the countryside compared to a long time ago, when there was a big outflow. This return has been caused by the fact that the big cities are finding it difficult to find work. Some of his employees with a degree in mechanics used to earn a salary of around R$3,000.00 in industry. Today they help with milk production and earn around R$2,000.00. They have tried to find a new opportunity in the big cities, but the companies only offer a salary of R$1,500.00 (WILL, 2017).

4.7.3 Grahl property

The property is located in the municipality of Braço do Trombudo, in the town of Ribeirao Ernesto, approximately 4 km from the city center. The property covers 11 hectares and is used for grazing cows for milk production, silage production and vegetables for consumption. The owner is Josely Eger Grahl (37), her mother Dagmar Eger (55) and her son Carlos Grahl (16) help with the family's milk production. Josely's husband works in an industry in the town (JOSELY EGER GRAHL, 2017).

Figure 15 - Josely, Dagmar and Carlos Grahl

Source: Author's collection (2017).

According to Grahl (2017), production was started in succession over many years by his mother. Before taking up dairy farming, they used to grow tobacco, and because they were always looking to increase productivity, they ended up opting for dairy farming, as it was considered a practical method and a job that was appreciated by the family. Table 08 below shows some information about the Grahl property.

Table 08 - Grahl property data

Milk Purchasing Company	N° Cows	No. of milkings	Approximate daily production
Tyrol	13	3	160

Source: Prepared by the author based on the interview (2017).

The family intends to stay in the business because they have been doing it for many years and because they are better suited to it. The negative points raised by the family are the changes in the price paid to the producer, along with the price of inputs to feed the animals. They currently receive R$0.80 per liter, and pointed out the reduction in the price of milk sales, however, there has been an increase in input costs, showing discontent with the valuation of the activity (GRAHL, 2017).

According to Grahl (2017), the monthly income comes from milk production, her mother's pension, and her husband who works outside the farm. Not all the milk produced is sold only to the dairy, but to buyers in the municipality.

Figure 16 - Grahl family property

Source: Author's collection (2017).

According to Grahl (2017), no member of the family currently has a degree, but his son Carlos intends to take over the business and study agronomy.

As for technology, in addition to the improvement of mechanical milking, refrigeration, which used to be done using teapots, where it took time to clean each one, today technology has contributed significantly to optimizing time in agriculture. Currently, they only use a few technological methods on their property, including animal insemination, which is carried out on the cows by a municipal employee, and pasture picketing (GRAHL, 2017).

Figure 17 - Dairy cattle on the Grahl property

Source: Author's collection (2017).

For the family, if the conditions of employment continue to be difficult, the young population that has migrated to the cities will be forced to return to the countryside in order to support themselves. There have been several cases of large numbers of employees being laid off by companies in the municipality and surrounding areas, and those who have a connection with agriculture will have an activity to secure themselves, if they don't have another activity, it will become difficult, as has been seen in recent years (GRAHL, 2017).

4.7.4 Vanderlinde property

The property is located in the municipality of Rio do Oeste, about 4 km from the city center. It covers 48 hectares, where in addition to pasture, rice and corn are produced. The family is made up of Renato Vanderlinde (65), Izete Vanderlinde (53), Douglas Vanderlinde (30), Daniel Vanderlinde (30) and their daughter Deise Vanderlinde (27).

Figure 18 - Daniel, Izete, Douglas and Renato Vanderlinde

Source: Author's collection (2017).

For Vanderlinde (2017), dairy farming has been carried out by the family for a long time, before Izete was born. Rice production is classified as a second alternative, since it is possible to produce milk regardless of the climate, while for rice, the floods in the region prevent high productivity. Corn is used to produce silage to feed the cows. Figure 18 below shows relevant information about the property.

Table 09 - Vanderlinde property data

Milk Purchasing Company	N° Cows	No. of milkings	Approximate daily production
Riolat	26	3	500

Source: Prepared by the author based on the interview (2017).

Dairy farming demands too much of the producer to keep it up to the standards required by the buying dairy, as it requires various chemical products, washing that meets these requirements, and maintaining the right amount of bacteria. The frequent changes in the price of milk make it impossible for producers to invest, and they can't plan financially with the price of milk they currently earn. The Vanderlinde family currently sells milk for R$1.00 a liter (VANDERLINDE, 2017).

Even in the face of setbacks in dairy farming, the family is gradually expanding its production by raising some calves that are in the growth phase, with more genetically advanced cattle. It's a long-term investment, planned for a year from now, where these animals will contribute to doubling productivity (VANDERLINDE, 2017).

According to Vanderlinde (2017), his sons work outside the property in a dairy in the town, but they help with production when they return. Renato worked for a company in Rio do Sul for around 32 years, but ended up returning to dairy farming because his wife Izete handled it alone. Not all the milk produced is sold to the dairy, but it is also sold to buyers in Rio do Oeste, who already know the family.

Vanderlinde family

Figure 19 -

Source: Author's collection (2017).

According to Vanderlinde (2017), the two sons intend to succeed the activity on the property. His son Douglas has taken training courses in milk quality and legal milk production, both run by SENAR. He says that knowledge helps with technological advances and is important for managing the property; however, courses for producers should be held and disseminated more frequently, focusing on areas where genetic methods can be used, such as insemination, which would save the producer costs if he were to carry out the procedure himself.

Figure 20 - Certificate of course taken

SENAR
Santa Catarina

CERTIFICADO

O Serviço Nacional de Aprendizagem Rural - SENAR confere o presente Certificado ao(à) Sr.(a)

DOUGLAS VANDERLINDE

CPF nº 06086701970, por seu aproveitamento no(a) Treinamento em Qualidade do Leite nº da CBO 6131-15, realizado (a) em RIO DO OESTE, no periodo de 22/04/2014 a 23/04/2014, com carga horária de 16 horas.

Florianópolis, 1 de novembro 2017

Superintendente

Presidente do Conselho Administrativo

Source: Author's collection (2017)

With technology, many of the conventional methods have become extinct, leading to a change in the way the animals are cleaned, with chemical products applied to the cow's teat during milking, reducing diseases and infections. The family has now adopted picketing, a refrigeration system and silage is also made by renting tractors (VANDERLINDE, 2017).

Figure 21 - Dairy cattle on the Vanderlinde property

Source: Author's collection (2017).

The fact that young people return to farming is only appropriate for those who already have some connection to the area through their parents. Young people may return because of the difficulties they are facing in the city at the moment, such as unemployment and low salaries. However, there are still young people looking for opportunities in industry and leaving agriculture (VANDERLINDE, 2017).

4.7.5 Norenberg Property

The property is located in Trombudo Central, about 8 km from the center of the municipality. The property covers 16 hectares, where in addition to pasture for dairy cows, it also produces sheep and corn for silage. The family is made up of Laércio Sydnei Norenberg (45 years old), Raquel Norenberg (43 years old) and Lyon Norenberg (16 years old), with only the couple working directly on the current production of the property (NORENBERG, 2017).

Figure 22 - Raquel and Laércio Norenberg

Source: Author's collection (2017).

According to Norenberg (2017), before producing milk, the family used to produce cassava, which not only cost a lot, but also didn't make much of a profit and the income was only annual. In 2000, as the property had a structure for dairy cattle, they invested in milking and in buying cows, and since then they have been working with this activity, which corresponds to a monthly income. They also produce sheep. Table 10 shows some of the property's data.

Table 10 - Norenberg property data

Milk Purchasing Company	N° Cows	No. of milkings	Approximate daily production
Edulat	25	3	235 liters

Source: Prepared by the author based on the interview (2017).

For Norenberg (2017), producing dairy cattle on pasture is more viable than feeding them on feed and silo. His property produces its own silo for the whole year, but the animal requires about 10% of its weight in feed, which can be purchased from the pasture, but with the production and extraction of milk, the animal needs more nutrients to complement the feed and silo. However, the cost of sacks of corn, fertilizer and especially diesel oil is extremely high, and without any subsidies from the state, the producer has to allocate himself to more affordable production. What the producer receives doesn't always really keep him in production.

As for the price, an agreement has been reached with the dairy that during the year the milk to be sold will remain at the price of R$1.10. Even when the price fluctuates more or less, it remains stable. The biggest obstacle in agricultural work is not having a fixed income, which makes it difficult for producers to control their budgets, both to maintain themselves and to invest (NORENBERG, 2017).

Among the other difficulties mentioned by the family is the climate, as the region has a lot of bad weather which affects the animal's feed for silo production. The animal gradually loses weight and then stops producing milk, where many cows produce a lot in one period, but then their udders become inflamed, resulting in mastitis (NORENBERG, 2017).

The family supports itself financially by producing milk and sheep, and a few liters of the milk produced are used to make cheese and cream to sell in the city center, thus supplementing their monthly income (NORENBERG, 2017).

Figure 23 - Norenberg family property

Source: Author's collection (2017).

The owner, Laércio, has a degree in Physical Education from the University of Contestado and taught for nine years after graduating. Even though he worked outside the property, he was always involved in agriculture, returning later to take over the family business, as well as being used to dealing with life in agriculture. Any instruction and information can be assimilated into everyday life to make it easier, especially when it comes to mechanizing agriculture. Today, it's necessary to acquire implements that can boost production and reduce labour (NORENBERG, 2017).

Figure 24 - Training certificate

Universidade do Contestado - UnC

O Reitor da Universidade do Contestado no uso de suas atribuições
e tendo em vista a conclusão do curso de

EDUCAÇÃO FÍSICA

em 16 de fevereiro de 2009, confere o título de
Licenciado em Educação Física a

LAÉRCIO SIDNEY NÖRENBERG

Carteira de Identidade nº 2.627.534-1
Nascido em 5 de novembro de 1972
Natural de Trombudo Central - SC, de nacionalidade brasileira
e outorga-lhe o presente Diploma, a fim de que possa gozar de todos os direitos
e prerrogativas legais.

Curitibanos - SC, 2 de abril de 2009.

Reitor

Diplomado

Carlos Eduardo Carvalho
Diretor Acadêmico

Source: Author's collection (2017).

In order to succeed to the property, the family has a son who has not yet chosen whether or not he will succeed to the activity. Regardless of his choice, his parents will provide him with all the knowledge he needs to succeed, and for him, technology and information will be much more beneficial to productivity (NORENBERG, 2017).

Technology has introduced many possible techniques to make it easier for producers. In the past, 15 to 20 people used to make the silo to feed the cows, and the cutting work was done by hand. Today, the silo is made by two people, and about 20 times more is cut with mechanization, because the silage maker is hydraulic, and the trailer is hydraulic to attach it to the tractor. What's more, it's reduced the working time from two weeks to two days. The electric fence also changed the material, where it used to take almost four days to picket the land. Today, with the more malleable plastic wire, it can be done in three hours with a screwdriver (NORENBERG, 2017).

Figure 25 - Nitrogen tank for storing bull semen

Source: Author's collection (2017).

According to Norenberg (2017), all the silage production and machinery is his own, and he applies and preserves semen for insemination in the cows. In order not to become outdated, technology must be obtained in the rural environment, because the property is like a company, where the production process is improved.

Figure 26 - Dairy cattle on the Norenberg property

Source: Author's collection (2017).

According to Norenberg (2017), the young population is returning to agriculture because food is now very industrialized and we no longer get the quality we did 30 years ago. And even though today's population has fewer children, these parents are part of a family with many siblings. In this way, the population has indirectly increased in recent years, and due to the exodus on the way to comfort and improvement in the city, food will be scarce for the sustainability of the population, so there must be a return to planting in agriculture, as a method of subsistence.

4.8 RESULTS OBTAINED

According to the data in Table 02, which shows the milk productivity reported by the families, it can be seen that four of them are involved in the activity because they have a family connection, i.e. through family succession, and only one is not. Farmer Laércio Norenberg pointed out that he started dairy farming on his own, although as a child his parents were involved in it, but he himself invested and started dairy farming (LAÉRCIO NORENBERG, 2017).

When asked whether they would stay in the business, according to Table 02, four of the interviewees said that they would stay in dairy farming because it is an activity with a monthly income and because they enjoy dealing with it. As for the interviewee who disagreed, it was because of the daily commitment involved in milk production, justifying the fact that dairy farming currently only serves to support the family on a monthly basis and not like a few years ago, when they acquired goods and because they intend to replace it with another productive area, in which it is less exhausting (LAÉRCIO NORENBERG, 2017).

Table 02 - Milk productivity

Milk productivity	Yes	No
Verified family involvement in production	4	1
Verified the permanence of the activity	4	1
Noticed attractiveness in production	4	1
Noticed difficulties in production	5	0

Source: prepared by the author based on the interviews (2017).

With regard to the attractiveness of production, four said they were positive about the return on production, but like any other activity, it should be developed with caution in order to acquire resources. The remaining family said it was not totally unfavorable, but it could be an activity with greater reciprocity in terms of price per liter sold (JOSELY EGER GRAHL, 2017).

All the families cited difficulties in dealing with production, such as price fluctuations in the milk sold and in the inputs used to feed the dairy cattle, which prevent them from making any long-term investments. They also noted the hygiene and health controls required by dairies, especially the Somatic Cell Count (SCC), which has changed in recent years, with rates reported as difficult to achieve.

With regard to some kind of academic training, according to table 03 below, of the five interviewees, four have degrees, technicians and training from agricultural organizations. The only remaining interviewee does not have any training, but one family member has the prospect of specializing in agriculture (JOSELY EGER GRALH, 2017).

Of the families interviewed, four said that the technology applied to milk production has been very significant. Josely Eger Grahl says: *"Yes, certainly even the cooler, which used to be a lot of work and now, in the time it took to wash a teapot, it washes the whole cooler." The* cleaning and sanitizing procedure has also changed and there has been a reduction in manpower. On the other hand, one family said that they had not adapted, claiming that they did not have the purchasing power to notice the benefits (MARCELO, 2017).

According to the interviewees, four families commented that knowledge is beneficial for carrying out agricultural production. Farmer Laércio Norenberg said that *"[...] all the information you absorb and assimilate you can use in your daily life. Because today, with the internet, even in terms of what you learned at university, [...] you look for information and it's very good quality"*. Still on the subject of knowledge in agriculture, Cristiano Will says that agricultural school gave him a basic overview, but with different notions from the other cattle farmers. Only one family said no, because they didn't have any training (GRAHL, 2017).

As for the application of knowledge in agricultural technology, three of the interviewees responded positively, according to Jeferson *Marcelo: "Yes, more and more technology is coming from pasture, animal welfare, which shows that it gives results. So much so that we're going to picket everything up here, to have more production, without a higher cost."* Still on the subject of technology, Laércio Norenberg says that ". *In the old days you used to make a silo [...] you'd go out to the fields with a knife and there'd be an average group of 15 to 20 people [...] today, I make a silo in two, me and one other. And so, I cut twenty times more than they did with mechanization*". For the other two who said no, one has no training and Douglas Vanderlinde says *"[...] we'd have to have more genetics, we'd have to have more knowledge", in other words,* more specialization focused on their production.

Of the families interviewed, all commented that they use some technological method on their property, in other words, that they use technology in favor of their production. Among those interviewed, Cristiano Will deals with "*artificial insemination, silo made from fodder, electronic milk collection system, composting system*". Most of the families use mechanized milking, which has long been an indispensable method, picketing, refrigerators, electric fencing made of soft plastic, insemination, either their own or carried out by the local town hall, silage harvesters and tractors, some rented and some owned.

Table 03 - Knowledge and technology

Knowledge and technology	Yes	No
Did you have any academic training?	4	1
Verified technology for the benefit of milk production	4	1
Verified knowledge as a benefit in agriculture	4	1
Verified application of knowledge in agricultural technology	3	2
Verified use of technology on the property	5	0

Source: prepared by the author based on the interviews (2017).

In view of the prospects for family succession, three of the interviewees claim that their children will be responsible for production when they leave the property, because it is an activity that the family has been involved in for many years and some of them currently work outside the property just to contribute to the family income. Laércio Norenberg adds that "[...] *we went to* do the *cachaça course in Sao Miguel together, but I leave it open to him if he wants to stay I'll encourage him, if he wants to study I'll encourage him, if he wants to have a simpler job, I'll support him*". Similarly, another family says that their daughter is not yet old enough to make such a decision (WILL, 2017).

Table 04 - Prospects for milk production

Prospects for milk production	Yes	No
Verified young family succession	3	2
Verified the prospects for the return of the young rural population	4	1

Source: prepared by the author based on the interviews (2017).

As for the prospects of young people returning to farming, four of the families interviewed believe that they will. According to Laércio Norenberg, *"[...] a lot of people are going to go back to farming [...] there will be a lack of food to sustain themselves, to maintain this comfort in the city [...] the return of these more educated people to farming will greatly improve the question of technology"*. Many believe in this assumption, because employability conditions are precarious at a national level, so many unemployed people who have links with agriculture return to it as a means of supporting themselves. Among other things, the salary range currently offered in industry is not so attractive compared to that offered in farming. As for the family who answered in the negative, they say that the exodus of young people from the countryside is still very permanent, according to Douglas Vanderlinde: *"Maybe it is, because of the difficulties encountered in the city [...], but for the time being, I see many young people from the countryside leaving in search of new opportunities [...]". The* employment issue could affect young people's participation in agriculture in the future.

5. CONCLUSION

Family farming plays a major role in the economic and social development of our state, along with other Brazilian regions. In this sense, family work has a strong link to develop new producers and generations to continue this productive process. Agriculture has grown over the last few years, becoming a target for study in academic circles, due to its cultural potential, the permanence and appreciation of rural areas, which has incorporated technological innovation through the knowledge and training of producers, in order to favor the production process, as in the case of milk production.

Technology in favor of agriculture, specifically family farming, has benefited through easier handling, reduced labor, optimized production on the property, thus resulting in less costly production. In view of this scenario, the aim of this study was to analyze the prospects for family farming in milk production in the municipalities of Atalanta, Agrolândia, Braço do Trombudo, Rio do Oeste and Trombudo Central. These prospects are based on the use of technology in rural areas, observing the link between academic education in favor of this use, family succession on the part of families and their point of view on the return of the young rural population.

With regard to the research objectives, the first was to gather information through an unstructured interview for the analysis of each municipality, which was applied to obtain the results relevant to the research.

The second objective was to demonstrate the use of technological innovation and training to benefit milk production. Among the families interviewed, it was found that all of them had used some kind of technological method in their production. The use of technology has proved to be beneficial, as it has considerably reduced labor costs, as well as reducing the time spent on the activity and the costs previously incurred in milk production. With regard to training in agriculture, most families had a member with some kind of education, including a degree, technical courses and training courses offered by rural institutions. Among the knowledge acquired, it was found that academic training has brought benefits to farming, as it has made it easier to use the technologies currently available.

The third objective was to diagnose family succession on the farm and prospects for the return of the young rural population to farming. The majority of the children interviewed said that they would succeed to the farm and maintain agricultural production, in this case milk production. As for the young people's prospects of returning, the families claim that the conditions of employment and unsatisfactory remuneration that currently come from the city are contributing to the return of this young rural population that already has a link with agriculture, serving as a means of support and financial stability. In fact, farming is proving to be a profitable sector that is growing in a promising way.

As explained above, the three objectives of the work were achieved, so it stands out that the families in the municipalities of Atalanta, Agrolândia, Braço do Trombudo, Rio do Oeste and Trombudo Central have followed technological methods, as well as some kind of academic training which, when applied to agriculture, have brought notable benefits to milk production. In other words, knowledge and technology have favored the production process in rural areas for families.

It is clear that in order to continue farming and producing milk in families, the children will be the ones to take over. It is estimated that young people will return to work in the fields, contributing to the notable growth of the agricultural sector.

Although milk production is covered by technology, there are still some difficulties for farmers, such as price fluctuations in milk and cattle feed, the sanitary and hygienic controls required, which are difficult to achieve, and the devaluation of the suggested price of a liter of milk, as well as the activity of producing it.

However, the subject presented in this work has not been exhausted. The subject is open to further research in the area, which could further promote the development of agriculture, milk production and rural areas, adding more knowledge and technology for family farmers in the municipalities presented and other regions. Once again, the importance of knowledge for the use of technological methods to optimize and strengthen production is highlighted, as well as encouraging young people to return to farming.

REFERENCES

ABRAMOVAY, Ricardo. **Family Farming and Land Use**. Sao Paulo, 1997. Available at: <http://produtos.seade.gov.br/produtos/spp/v11n02/v11n02_08.pdf>. Accessed on: May 20, 2017.

, R. **Family Farming and Territorial Development. Agrarian reform.** v. 28, n.1, p.2, 1999. Available at: <http://wp.ufpel.edu.br/ppgdtsa/files/2014/10/Texto-Abramovay-R.-Agricultura-familiar-e-desenvolvimento-territorial.pdf>. Accessed on May 29, 2017.

R. The social capital of territories: rethinking rural development. **Revista Economia aplicada**. n.2. vol. 5. 379-397, April/June 2000.

R; BALDISSERA, Ivan Tadeu; CORTINA, Nelson; FERRARI, Dilvan; SILVESTRO, Milton and TESTA, Vilson Marcos (1998). **Juventude e agricultura familiar: desafios dos novos padrões sucessórios** - Chapecó, Brasilia - FAO-INCRA/CPPP- EPAGRI Agreement - UNESCO Editions.

R. **Family farming and public service:** new challenges for rural extension. Cadernos de Ciência & Tecnologia, v.15, n.1.Brasilia: 1998.

ARBAGE, Alessandro Porporatti. **Fundamentals of Rural Economics**. Santa Catarina: Argos, 2006.

ATLAS OF HUMAN DEVELOPMENT IN BRAZIL. Consultation. Available at: <http://www.atlasbrasil.org.br/2013/pt/consulta/>. Accessed on May 20, 2017.

AZEVEDO, Francisco Fransualdo; PESSÔA, Vera Lùcia Salazar. **The National Program to Strengthen Family Farming in Brazil:** an analysis of the regional and sectoral distribution of resources. Sociedade & Natureza, Uberlândia, ano, v. 23, p. 483-496, 2011. Available at:< http://www.scielo.br/pdf/sn/v23n3/v23n3a09.pdf>. Accessed on: May 20, 2017.

BANCO DO BRASIL. Inter-American Institute for Cooperation on Agriculture. Dairy farming. Sustainable regional development. v. 1. Brasilia: Banco do Brasil Foundation; 2010. **Series of Proposals for Action in Production Chains.** Available at <http://www.bb.com.br/docs/pub/inst/dwn/Vol1BovinoLeite.pdf>. Accessed on June 25, 2017.

BRAZIL. Law No. 11.326, of July 24, 2006 of the Presidency of the Republic. Establishes the Guidelines for the Formulation of the National Policy for Family Farming and Rural Family Enterprises . Available at: <http://www.planalto.gov.br/ccivil_03/_ato2004-2006/2006/lei/l11326.htm>. Accessed on: May 20,

2017.

BRAZIL, P. N. U. D. United Nations Development Program.

Human Development and HDI. Available at: <http://www.pnud.org.br/IDH/DH.aspx> . Accessed on: August 18, 2017.

BONAT, Débora. **Research Methodology.** 3 ed. Curitiba: IESDE, 2009.

COOPERATIVA SANTA CLARA. **CCS and CBT results by message benefit milk quality**. Available at: < http://www.coopsantaclara.com.br/noticias/277- results-of-ccs-and-cbt-by-message-benefit-the-quality-of-milk >. Accessed on October 20, 2017.

DIARIO CATARINENSE. **Technology in milk production guarantees higher quality and better prices in Santa Catarina.** Available at <http://dc.clicrbs.com.br/sc/noticia/2010/11/tecnologia-na-producao-de-leite-garante-maior-qualidade-e-melhor-preco-em-santa-catarina-3109657.html>. Accessed on October 18, 2017.

DOS SANTOS, Adriana BA; FAZION, Cintia b.; DE MEROE, Giuliano PS. Innovation: A Study on the Evolution of Schumpeter's Concept. **Caderno de Administraçao**. Revista da Faculdade de Administraçao da fea. issn 1414-7394, v. 5, n. 1, 2011.

BRAZILIAN AGRICULTURAL RESEARCH Corporation - EMBRAPA. Milk agribusiness . Available at:<http://www.agencia.cnptia.embrapa.br/Agencia8/AG01/arvore/AG01_191_21720039246 .html> Accessed October 20, 2017.

FROEHLICH, JM; et al. Selective exodus, masculinization and aging of the rural population in the central region of RS. **Ciência Rural**. 41, 9, 1674-1680, September. 2011.

GIL, Antônio Carlos, 2002. **Como Elaborar Projetos de Pesquisa**. 4ed. Atlas, 2002.

GRANATO, Daniel; NUNES, Domingos Sàvio. **Chemical analysis, functional properties and quality control of foods and beverages**. 1ed. Rio de Janeiro: Elsevier Editora Ltda, 2016.

BRAZILIAN INSTITUTE OF GEOGRAPHY AND STATISTICS - IBGE Cities. Santa Catherine. Available at: <https://cidades.ibge.gov.br/xtras/uf.php?lang=&coduf=42&search=santa-catarina> Accessed on: August 14, 2017.

. IBGECidades . Atalanta. Available at: <https://cidades.ibge.gov.br/xtras/perfil.php?lang=&codmun=420180&search=santa-catarina|atalanta>. Accessed on: August 14, 2017a.

. IBGE Cities. History of Atalanta. Available at: <http://www.cidades.ibge.gov.br/painel/historico.php?lang=&codmun=420180&search=santa - catarina|atalanta|infograficos:-historico>. Accessed on: August 14, 2017b.

. IBGECities . Agrolandia Available at: <https://cidades.ibge.gov.br/xtras/perfil.php?lang=&codmun=420020&search=santa-catarina|agrolandia>. Accessed on August 14, 2017c.

. IBGE Cities. Agrolândia History. Available at: <https://cidades.ibge.gov.br/painel/historico.php?lang=&codmun=420020&search=santa-catarina|agrolandia|infograficos:-historico>. Accessed on August 14, 2017d.

. IBGE Cities. Braço do Trombudo. Available at: <https://cidades.ibge.gov.br/xtras/perfil.php?lang=&codmun=420285&search=santa-catarina|braco-do-trombudo>. Accessed on August 14, 2017e.

IBGE Cities. Rio do Oeste. Available at: <https://cidades.ibge.gov.br/xtras/perfil.php?lang=&codmun=421460&search=santa- catarina|rio-do-oeste>. Accessed on August 14, 2017f.

. IBGE Cities. History of Rio do Oeste. Available at: <https://cidades.ibge.gov.br/painel/historico.php?lang=&codmun=421460&search=santa-catarina|rio-do-oeste|infograficos:-historico>. Accessed on August 14, 2017g.

. IBGE Cities. Trombudo Central. Available at: <https://cidades.ibge.gov.br/xtras/perfil.php?lang=&codmun=421860&search=santa-catarina|trombudo-central>. Accessed on August 14, 2017h.

. IBGE Cities. History of Trombudo Central. Available at: <https://cidades.ibge.gov.br/painel/historico.php?lang=&codmun=421860&search=santa-catarina|trombudo-central>. Accessed on August 14, 2017i.

LAKATOS, Eva Maria. MARCONI, Marina de Andrade. **Fundamentals of Scientific Methodology.** 5 ed. Sao Paulo: Atlas, 2003.

MÜLLER, Ernst Eckehardt. Milk quality, somatic cells and mastitis prevention. **Symposium on the Sustainability of Dairy Farming in Southern Brazil**, v. 2, n. 2002, p. 206-217, 2002.

NOVINSKI, Charles Ortiz. **Corn silage production on small rural properties.** 2013. Available at: <http://www.ensilagem.com.br/producao-de-silagem-de-milho- on-small-rural-properties/>. Accessed on October 20, 2017.

UNITED NATIONS FOOD AND AGRICULTURE ORGANIZATION - FAO (RomeItaly) (Org.) What is family farming? 2014. Available at: http://www.fao.org/family-farming-2014/home/what-is-family-farming/pt/. Accessed on: May 12, 2017.

AGRICULTURAL POLICY. **The persistent rural-urban migration**. Year XVIII. n. 4, Oct/Nov/Dec. 2009. Available at: <file:///C:/Users/39494/Downloads/Revista%20de%20Pol%C3%ADtica%20Agr%C3%ADcola%20n4-2009.pdf> Accessed May 15, 2017.

ATALANTA CITY HALL, 2017. Tourism. Available at: <http://atalanta.sc.gov.br/turismo/>. Accessed on August 14, 2017.

AGROLÂNDIA CITY HALL, 2017. Tourism. Available at: <http://agrolandia.sc.gov.br/turismo/informacoes/>. Accessed on August 21, 2017.

MUNICIPALITY OF BRAÇO DO TROMBUDO, 2017. Tourism. Available at:< http://bracodotrombudo.sc.gov.br/turismo/>. Accessed on August 22, 2017.

MUNICIPALITY OF RIO DO OESTE, 2017. History. Available at: <http://www.riodooeste.sc.gov.br/cms/link/link-cabecalho/codMapaItem/83415>. Accessed on August 23, 2017.

Municipality OF TROMBUDO CENTRAL, 2017. Tourism. Available at: < http://trombudocentral.sc.gov.br/turismo/item/Atrativos/>. Accessed on August 28, 2017.

PROCHNIK, Victor. Production chains and industrial complexes. **Hasenclever, L.; Kluper, D. Industrial Organization.** Rio de Janeiro: Campus, 2002.

SANTOS, OV dos; MARCONDES, T.; CORDEIRO, !FL. Study of the milk chain in Santa Catarina: prospects and demands. **Florianópolis:** Epagri, 2007.

SENAR, Precision farming. **Technological advances improve productivity in agriculture**. 2016. Available at: < http://www.senar.org.br/agricultura-precisao/avancos- tecnologicos-melhoraram-a-produtividade-na-agricultura/>. Accessed on June 20, 2017.

SCHMITZ, Heribert. **Family farming:** rural extension and participatory research. Sao Paulo: Annablume, 2010.

SCHNEIDER, Carlise Porto. Family farming and entrepreneurship: a study on the trajectories of young people who graduated from the Rural Youth Development Center (Cedejor) in Vale do Rio Pardo/RS. 2008.

SCHNEIDER, Sérgio. **Social Theory, Family Farming and Pluriactivity.** Brazilian Journal of

Social Sciences. Vol - 18. February/2003.

SILVA, José Graziliano da; GROSSI, Mauro Eduardo Del. **The New Brazilian Countryside**. Sao Paulo: Unicamp, 1998. Available at at: <http://www.iapar.br/arquivos/File/zip_pdf/novo_rural_br.pdf>. Accessed on: June 31, 2017.

SILVA, Luis César. Agricultural Products Production Chain. Federal University of Espirito Santo: **Department of Rural Engineering**. Technical Bulletin: MS: 01/05 on April 21, 2005 (2005).

SILVA, Rosana de Oliveira Pithan e. On the New Normative Instruction n. 7 for Milk Quality. **Institute of Agricultural Economics.** V. 11, July 2016.

SOUZA, Nali de Jesus de. **Economic development**. 5. ed. Sao Paulo: Atlas, 2011.

TINOCO, Sonia Terezinha Julianatto. **Conceptualization of family farming**: a bibliographical review. Dextru - Rural Extension Division, Sao Paulo: Unesp, 2006. Available at: <http://www.infobibos.com/Artigos/2008_4/AgricFamiliar/Index.htm>. Accessed on: May 20, 2017.

TOMEI, Patricia Amelia; LIMA, Daniela Alves. The rural entrepreneur and innovation in the Brazilian context. In: **Anais... XI National Congress of Excellence in Management**. 2015.

VIANA, Giomar; FERRAS, Robson Paulo Ribeiro. The milk production chain: a study on the organization of the chain and its importance for regional development. **Revista Capital Cientifico-Eletrônica (RCOe)-ISSN 2177-4153**, v. 5, n. 1, p. 23-40, 2010.

VILCKAS, Mariângela. NANTES, José Flàvio Diniz. **Adding Value:** An Alternative for the Expansion of the Organic Food Market. Rural & Agro-Industrial Organizations, Lavras, 2007. Available at:<http://ageconsearch.umn.edu/bitstream/43728/2/(02)%20Article%2007.290.pdf>. Accessed on: June 3, 2017.

ZANELLA, Cleunice; LEITE, André Luis da Silva. Innovation in the poultry production chain: a case study in an agro-industry in the state of Santa Catarina. **Rural & Agroindustrial Organizations**. 18, 2, 186-201, June 2016. ISSN: 15173879.

APPENDIX 1: INTERVIEW WITH DAIRY FARMERS

Identification of interviewee(s):

Name of owners:

Age:

Resident municipality:

Interview duration:

CHARACTERIZATION OF THE PROPERTY.	Property size: A company that buys milk: Number of cows: Number of milkings: Other production besides milk:
MILK PRODUCTIVITY	1 - Why did you choose dairy farming? 2 - Have you always worked with this production? 3 - At the moment, is milk production attractive to farmers? 4 - Do you intend to stay in dairy farming? 5 - What are the main difficulties encountered in milk production?
KNOWLEDGE AND TECHNOLOGY	1 - Do you have a degree or diploma? 2 - With your experience in production, is the knowledge you have acquired necessary to relate to technological advances and innovations over the years? 3 - Have there been any notable benefits in terms of agricultural technology in your production? 4 - Among the technological methods that have emerged in dairy farming, do you apply any on your property?
PROSPECTS FOR MILK PRODUCTION	1 - If you want to continue producing milk on your property in the future, is there any prospect of family succession? 2 - There have been many cases in which young people have left the countryside in search of better opportunities in the job market, however, there are a few cases in which they return to the farm as graduates with some education. In your opinion, will this process of returning to the farm grow over the years?

Source: own elaboration (2017).

MIX
Papier aus verantwortungsvollen Quellen
Paper from responsible sources
FSC® C105338

Printed by Books on Demand GmbH, Norderstedt / Germany